The Checkbook Series

Microelectronic Systems N2 Checkbook

R E Vears

D1147301

NEWNES

An imprint of Butterworth-Heinemann Ltd

Newnes
An imprint of Butterworth-Heinemann Ltd
Halley Court, Jordan Hill, Oxford OX2 8EJ

 PART OF REED INTERNATIONAL BOOKS

OXFORD   LONDON   GUILDFORD   BOSTON
MUNICH   NEW DELHI   SINGAPORE   SYDNEY
TOKYO   TORONTO   WELLINGTON

First published 1982
Reprinted 1985, 1986
Second edition 1988
Reprinted 1990, 1991

British Library Cataloguing in Publication Data
Vears, R. E.
    Microelectronic Systems N2 Checkbook
    1.   Microcomputers   2. Microprocessors
    I.   Title
    621.3819'58   TK7888.3

ISBN 0 7506 0160 4

Printed and bound in Great Britain by
Hartnolls Ltd, Bodmin, Cornwall

Contents

Note to readers

Checkbooks are designed for students seeking technician or equivalent qualification through the courses of the Business and Technician Education Council (BTEC), the Scottish Technical Education Council, Australian Technical and Further Education Departments, East and West African Examinations Council and other comparable examining authorities in technical subjects.

   Checkbooks use problems and worked examples to establish and exemplify the theory contained in technical syllabuses. *Checkbook* readers gain real understanding through seeing problems solved and through solving problems themselves. *Checkbooks* do not supplant fuller textbooks, but rather supplement them with an alternative emphasis and an ample provision of worked and unworked problems, essential data, short answer and multi-choice questions (with answers where possible).

Preface

This textbook of worked problems provides coverage of the Business and Technician Education Council level NII unit in Microelectronic Systems (syllabus U86/333). However, it can be regarded as a textbook in microelectronic systems for a much wider range of studies.

The aim of this book is to provide a foundation in microelectronic systems hardware and software techniques. Each topic considered in the text is presented in a way that assumes in the reader only the knowledge attained in BTEC Information Technology Studies F, Engineering Fundamentals F, or equivalent. Additional material on the basic ideas of systems, logic functions and numbering systems is included for the sake of completeness.

This book concentrates on the highly popular 6502, Z80 and 6800 microprocessors and contains approximately 80 tested programs which may be used with little or no modification on most systems based on these microprocessors. The text includes over 140 worked problems followed by some 250 further problems.

The author would like to express his thanks to the general editors, J. O. Bird and A. J. C. May for their helpful advice and careful checking of the manuscript. Finally the author would like to add a special word of thanks to his wife Rosemary, for her patience and encouragement during the preparation of this book.

The publishers and author would also like to thank the following firms for permission to reproduce diagrams and data in this book – Intel; Zilog; MOS Technology Inc; Mostek UK Ltd; Motorola Semiconductor Products Inc.

R E Vears
Highbury College of Technology
Portsmouth

1 Basic ideas of systems

A MAIN POINTS CONCERNED WITH THE BASIC IDEAS OF SYSTEMS

1 A system is defined as an orderly arrangement of physical or abstract objects. Systems have **inputs** and **outputs** arranged as shown in *Figure 1*.

The input signal may cause the system output to change or may cause the operation of the system to change. Therefore, the input signal is the **cause** of the change. The action which occurs as a result of an input signal is called the **effect**. The response of the system to an input signal is called the **process**.

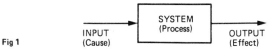

Fig 1

2 *Figure 2* shows the operation of a simple system. An input signal (voltage) causes the system to produce an output signal which is twice as large. Note that zero (0 V) input results in zero output. The output is always an enlarged version of the input signal, and the system is said to process the input signal. This particular system is called a **voltage amplifier**.

3 The system shown in *Figure 3* has three possible inputs, each of which may be connected to ground (0 V) via switch Sw. When each of the individual inputs is

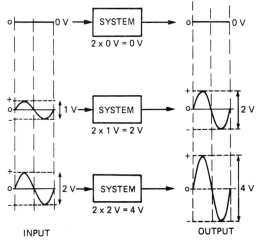

Figure 2

1

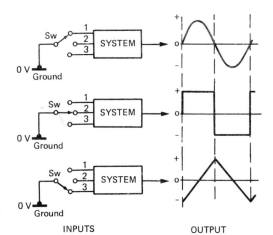

Figure 3   INPUTS   OUTPUT

connected to ground, a particular output signal shape results. To produce each different shape, a different process takes place within the system. Therefore in this case, the input signal is causing the system process to change. This particular system is sometimes called a **function generator**.

4 Examples of four other systems are shown in *Figure 4*.
The input, output and process of each of these systems are:

(a) Lamp dimmer system

input: variable voltage supplied from a manually operated rotary control.

output: variable intensity light.

process: adjust supply current to lamp according to the setting of the dimmer control.

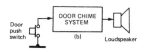

Figure 4(a)

(b) Door chime system

input: fixed voltage provided by manually-operated door switch.

output: a sequence of musical notes, or complete tunes.

process: upon receipt of an input signal, provide a suitable signal to drive the loudspeaker, and provide all timing for note pitch and duration.

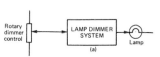

Figure 4(b)

(c) Coin changer system

input: coins.

output: coins.

process: determine value and validity of input coin(s) and compute amount of change to be given; may also be used in vending machines to calculate amount of change due when input coins result in overpayment.

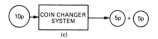

Figure 4(c)

(d) Oven control system

input: variable voltage from manually-adjusted temperature setting control and oven temperature sensing device.

output: heat at controlled temperature.

process: compare actual oven temperature with the desired (target) temperature and adjust the heating element current to maintain these two temperatures as close as possible to one another.

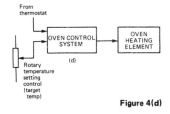

Figure 4(d)

5 In many systems, the input signal alone has insufficient power to operate the output device. In this case, an additional input to the system is required, that is, a power supply. The input signal then **controls** the flow of current between this additional input and the system output. One method by which this is achieved is shown in *Figure 5*.

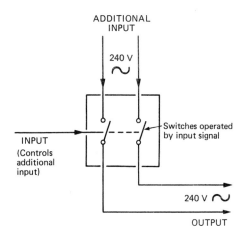

Figure 5

6 The input and output signals of a system are energy sources. An electronic system requires an input of electrical energy, but the input energy source is in all probability not electrical. The output from the system is in the form of electrical energy, but in most cases, the output required is some other form of energy, for example, heat, light or mechanical. Therefore, devices which are capable of converting energy from one form into another are an essential part of most systems. These devices are called **transducers**, and examples of typical transducers are illustrated in *Figures 6 and 7.*

A thermistor is a device often found in temperature measuring systems. It consists of a piece of special material to which two connecting wires are fixed, and has the characteristic that its electrical resistance changes according to its temperature. Two types of thermistor action are available:

3

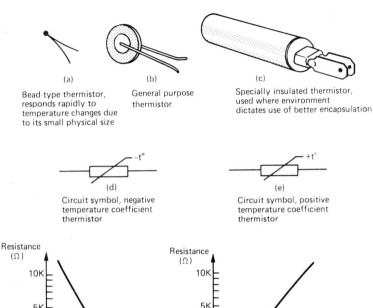

(a)

Bead-type thermistor, responds rapidly to temperature changes due to its small physical size

(b)

General purpose thermistor

(c)

Specially insulated thermistor, used where environment dictates use of better encapsulation

(d)

Circuit symbol, negative temperature coefficient thermistor

(e)

Circuit symbol, positive temperature coefficient thermistor

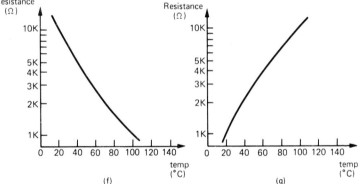

(f)

Characteristics of a negative temperature coefficient thermistor

(g)

Characteristics of a positive temperature coefficient thermistor

Figure 6

(a) increase in resistance as temperature increases, which is known as a **positive** temperature coefficient of resistance (PTC thermistor)

(b) reduction in resistance as temperature increases, which is known as a **negative** temperature coefficient of resistance (NTC thermistor)

Thus, the thermistor may be used as a transducer to convert heat energy into an equivalent electrical signal.

Some systems need to know when liquid in a container reaches a predetermined level. *Figure 7* shows two methods by which this may be accomplished. In *Figure 7(a)*, the rising liquid level causes the air pressure in the lower part of the transducer to increase. This, in turn, puts pressure on a spring-loaded diaphragm to which an electrical contact is fitted. At a predetermined pressure (and hence,

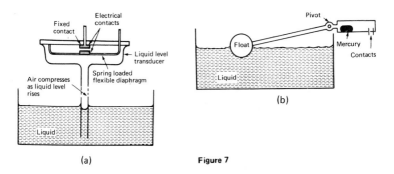

(a)

(b)

Figure 7

liquid level) the diaphragm suddenly springs across and causes the diaphragm to touch the fixed contact, thus completing the circuit to which it is connected.

An alternative method is shown in *Figure 7(b)* in which a small pool of mercury is used to complete the circuit between two fixed contacts. The contacts and mercury are housed in a small container, which is attached to a float and pivot assembly. The float rises with the liquid level, and at a predetermined level, the angle of the mercury container is such that the pool of mercury rolls down the container and makes an electrical connection between the two fixed contacts (note: mercury is a good conductor of electricity).

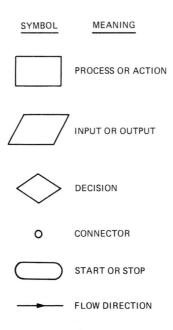

SYMBOL	MEANING
▭	PROCESS OR ACTION
▱	INPUT OR OUTPUT
◇	DECISION
O	CONNECTOR
⬭	START OR STOP
→	FLOW DIRECTION

Figure 8

7 The manner in which a system operates is determined by the functions performed by each of its blocks and by the sequence in which they are operated. This operating sequence may be represented by means of a **flow chart** which is drawn using symbols similar to those shown in *Figure 8*.

B WORKED PROBLEMS ON THE BASIC IDEAS OF SYSTEMS

Problem 1   What is the basic function of a photo-electric device? Describe, using diagrams, **three** practical applications for photo-electric devices in systems.

The basic function of a photo-electric device is to convert light energy into a corresponding electrical signal. There are several different types of photo-electric devices. Some types have an electrical resistance which varies according to the amount of light falling on them, whilst other types generate an electrical potential (voltage). Examples of applications of the photo-electric device in practical systems are as follows:

(a) Conveyor belt control   The conveyor belt drive motor operates until an object arrives to break the beam between the light source and the photo-electric device. The drive belt then stops, and work may be carried out on the stationary object. The belt drive may be restarted by several methods. Examples include manually moving the object clear of the beam, or electrical override of the system, or a timed halt period with automatic restart (see *Figure 9*).

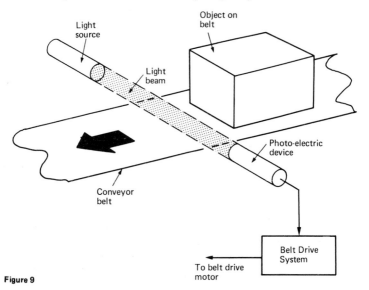

Figure 9

6

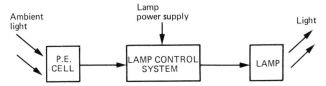

Figure 10

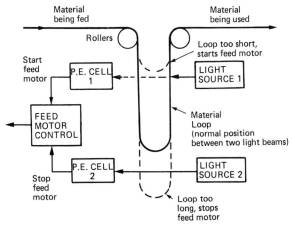

Figure 11

(b) Automatic lighting system   The photo-electric device (PE cell) detects when the ambient light level falls below a predetermined acceptable minimum, and causes the system to switch on the artificial illumination. The photo-electric device must be positioned so that it is not affected by the artificial illumination (see *Figure 10*).

(c) Control of feed rate for paper or similar sheet material   Where the material is used up at a variable rate, the feed drive motor needs to be carefully controlled (*Figure 11*). In this system, a loop of the material is allowed to sag between two light beams. When the lower beam is interrupted it is an indication that the feed rate is too great, and the feed drive must be stopped. When the upper beam is restored it is an indication that the feed drive must be switched on again. Thus the feed drive is controlled so that the length of material in the loop is such that it always hangs between the two light beams.

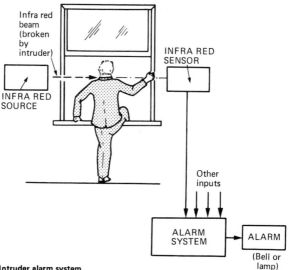

Figure 12   Intruder alarm system

Problem 2   With the aid of a sketch, show the principle of operation of an intruder alarm system. State the type of transducers used, and give reasons for your choice.

All vulnerable points in a building may be protected by infra-red beams, each beam being directed at an infra-red receiver connected to the alarm system. When a beam is broken by the body or limb of the intruder, this is detected by the alarm system and a warning signal is given. The warning may be a light or audible alarm in a security centre.

The system is usually arranged so that restoration of the infra-red beam does not shut off the alarm. Two transducers are required at each point to be protected, one to convert electrical energy into infra-red energy, and another to convert the received infra-red beam back into an electrical signal. A typical installation is shown in *Figure 12*.

Infra-red energy is identical in its characteristics to ordinary light except that it is invisible to the human eye. This means that an intruder will not be able to see the beams and thus avoid them. The infra-red receiver is similar to an ordinary photo-electric cell except that it is made sensitive to infra-red rather than visible light. The output transducer is either an electric bell/buzzer or a lamp.

Problem 3   Describe a typical transducer to convert each of the following physical quantities into an equivalent electrical signal: (a) rotary position of a shaft; (b) heat; (c) light; (d) mechanical strain.

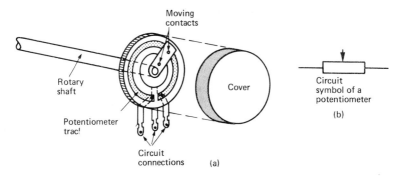

Figure 13 (a) Arrangement of a potentiometer (b) Circuit symbol

(a) A potential is applied across the ends of a potentiometer track which is made from carbon or some similar conductive material which has a suitable electrical resistance. A slider, in contact with the track, is rotated and a variable potential is obtained from the slider connection. A shaft is connected to the slider, therefore the position of the shaft determines the output potential (see *Figure 13*).

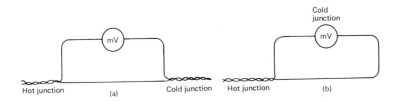

Figure 14   Typical connections

(b) If two dissimilar metals are connected as shown in *Figure 14(a)*, a potential difference (voltage) occurs if the two junctions are held at different temperatures. Suitable metals for the junctions are copper and constantan. Sometimes it is more convenient to make the cold junction the contact with the millivoltmeter or external circuit, as shown in *Figure 14(b)*. This device is known as a **thermocouple**.

(c) The photo-electric device illustrated in *Figure 15* has an electrical resistance which depends upon the intensity of light falling on it. The resistance is high under dark conditions, and falls as the illumination increases. Many materials exhibit this property, for example, cadmium sulphide or selenium, and they are thinly sprayed onto metallic foils, comb shaped so as to increase the effective contact length.

(d) Mechanical strain is a change in dimension of a material due to the application of an external force. It is often necessary to know how much strain there is in mechanical structures, especially when they are loaded. A **strain gauge** may be used as a transducer for this purpose, and a typical construction is shown in *Figure*

9

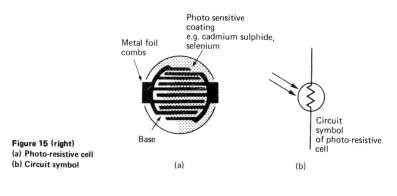

Photo sensitive
coating
e.g. cadmium sulphide,
selenium

Metal foil
combs

Circuit
symbol
of photo-resistive
cell

Base

Figure 15 (right)
(a) Photo-resistive cell
(b) Circuit symbol

(a)

(b)

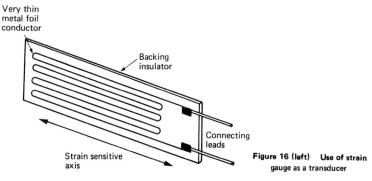

Very thin
metal foil
conductor

Backing
insulator

Connecting
leads

Strain sensitive
axis

Figure 16 (left)   Use of strain
gauge as a transducer

16. The strain gauge is fixed to the mechanical structure with a suitable adhesive, so that when the structure is subjected to strain the gauge is also strained. This causes a change in the electrical resistance of the fine conductors of the gauge which can be detected with suitable equipment to give a reading of strain.

Problem 4   An opto-isolator is often used to couple a signal between two parts of a system. Describe how this device functions and state its main advantage.

An opto-isolator (sometimes called a photo-coupler) consists of two transducers, one to convert an electrical signal into light, and a second one to convert the light back into an electrical signal. The signal is thus coupled between stages by means of a light beam only, and there is no direct electrical connection between the two stages of the system. The electrical insulation between the two transducers can be made to withstand very high potentials. This device, therefore, has the advantage that stages operating at high or very different potentials may be safely coupled without danger of electrical breakdown (see *Figure 17*).

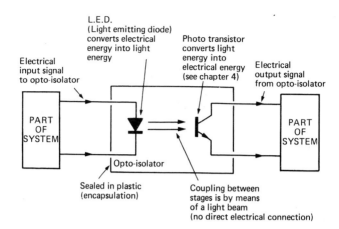

L.E.D.
(Light emitting diode)
converts electrical
energy into light
energy

Photo transistor
converts light
energy into
electrical energy
(see chapter 4)

Electrical
input signal
to opto-isolator

Electrical
output signal
from opto-isolator

PART
OF
SYSTEM

PART
OF
SYSTEM

Opto-isolator

Sealed in plastic
(encapsulation)

Coupling between
stages is by means
of a light beam
(no direct electrical connection)

Note: Opto-isolator is also called
a photo-coupler

Figure 17   Use of an opto-isolator

Problem 5 Draw a flow chart to show the operation of a 24 hour digital clock, having a one second delay mechanism. Show what modifications are necessary to convert the clock to operate on a 12 hour cycle.

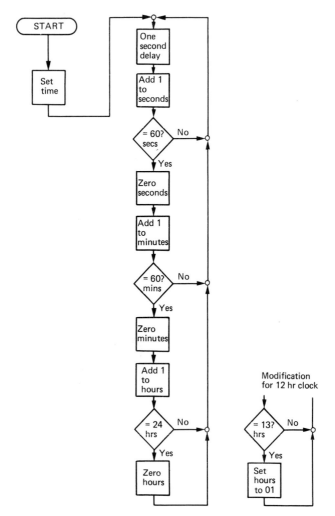

Figure 18

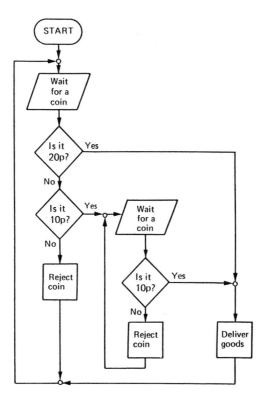

Figure 19

13

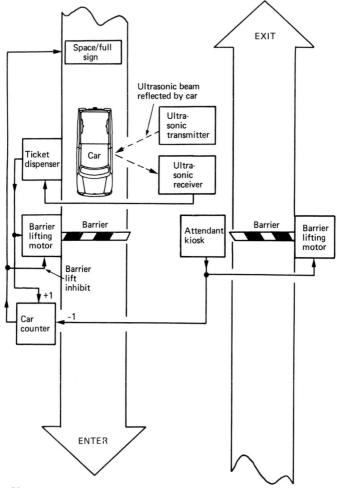

Figure 20

14

Problem 8   A mains operated heating element is used to heat a container full of liquid. The target temperature of the liquid is preset by means of a rotary control or a keyboard control, and the actual liquid temperature is shown on a digital display. Draw a block diagram of a suitable system for controlling the heater.

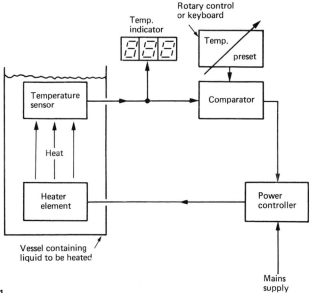

Figure 21

Problem 9   Draw the flow chart symbol used to represent a decision and explain fully the use of all of its possible outputs.

The decision symbol has one input and **three** possible outputs which are **'greater than'**, **'equal to'** and **'less than'**, as shown in *Figure 22*. To illustrate the use of all three outputs, consider the following example. A metal plate is etched so that its finished thickness is 0.1 mm. When the thickness is measured there are three possible results, thicker than 0.1 mm, exactly 0.1 mm and thinner than 0.1 mm. This is illustrated in *Figure 23* and according to the result obtained, the program flow will proceed down the appropriate branch from the decision box.

In many cases, however, there can only be two outputs from a decision box, **yes (true)** or **no (false)**, see *Figure 24*. Consider the case where a switch is tested to see if it is closed. Clearly the switch is either closed or not closed and these are the only two possibilities.

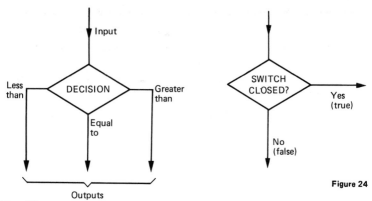

Figure 22

Figure 24

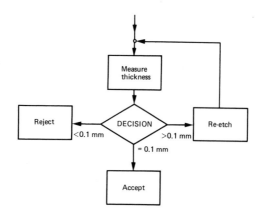

Figure 23

C FURTHER PROBLEMS ON THE BASIC IDEAS OF SYSTEMS

(a) SHORT ANSWER PROBLEMS

1 An orderly arrangement of physical or abstract objects is called a.............................

2 A signal which causes a system to react is called an..

3 A thermistor is a device which responds to changes in...

4 A device which is capable of converting energy from one form to another is called a

   ....

5 Input and output signals are sources of..

6 A photo-electric device responds to changes in........................   ....

7 An additional input to a system is required if it is necessary to convert an input

   signal into an..............................   ....

16

8 An device whose resistance falls with increase in temperature has a .... of resistance.

9 Sound vibrations which are above the normal range of human hearing are called .... signals.

10 A device which generates an electrical potential (voltage) when its two dissimilar metal junctions are held at different temperatures is called a ....

11 The sequence of operations which occur in a system is called a.................................

12 A graphical method of representing a sequence of operations in a system is called ....

13 What is the function of a controller in a system?

14 Draw the flow chart symbol which represents a decision.

15 List the conditions which are associated with the three outputs of a flow chart decision symbol.

16 The effect of changing the program in a system is to alter the....................................

17 What is meant by the term program flow?

18 The advantages of using a microelectronic controller rather than an electromechanical controller are...

(b) CONVENTIONAL PROBLEMS

1 With the aid of a simple diagram, describe fully what is meant by the term 'system'.

2 What is the fundamental purpose of a transducer? Describe the construction of **two** different types of transducer and state for what purpose each may be used.

3 With the aid of simple examples, show that a system may, (a) process its inputs to alter its output, or (b) process its inputs to alter its internal condition.

4 List **four** systems found in a typical home. For one of these systems, state its input(s), output(s) and process.

5 In many cases, a system input signal has insufficient power to control the system output. Describe how this problem is overcome in a practical system.

6 Describe a simple system for controlling the temperature in an electric oven, indicating any transducers used. Justify this as a system by listing the input(s), output(s) and process involved.

7 It is necessary to control the amount of smoke emission from the chimney of an industrial plant. Describe a simple method for measuring the smoke density in a chimney, stating the types of transducers used. Describe any special precautions you may think necessary with the transducers.

8 A supermarket check-out has a small conveyor belt system. Each time that the check-out operator removes an article from the belt to register the price, the belt moves forward to bring the next article within reach. Describe a simple method for controlling the movement of the belt, showing the types of transducers used.

9 The contents of a deep freeze cabinet are protected by an alarm system. If the temperature in the cabinet rises above the permitted value, an alarm sounds. Describe a suitable alarm system, indicating the types of transducers used.

10 Name **three** types of transducer which may be used to measure temperature. Describe the operation and characteristics of each type.

11 Draw a flow chart to show the sequence of events which occur when a car enters the car park illustrated in *Figure 20*.

12 Draw a flow chart to show the operation of the vehicle counter and the space/full display of the car park illustrated in *Figure 20*.

13 Boxes to be filled with small parts are placed on a conveyor belt. Each box is filled from a hopper with 100 parts, then the conveyor belt moves to bring the next empty box under the hopper. Draw a block diagram of a suitable system to perform this function.

14 Draw a flow chart to show the sequence of events in the conveyor belt system described in *Problem 13*.

15 With reference to the system illustrated in *Figure 21* draw a table to show the input, output and process of each block.

16 Draw a block diagram of a domestic washing machine and identify the input, output and process of each block.

17 List **five** systems found in a typical home which could be improved by using a microelectronic controller in their construction. Describe one of these systems in detail, and explain why you think that it is improved by being controlled by a microelectronic device.

18 Draw the block diagram of a typical domestic central heating system.

19 Write a program to illustrate a typical operating sequence for the central heating system described in *Problem 18*.

20 A room is protected against unauthorized entry by means of a security system. To gain entry to the room, a personal card is inserted into a slot and four digits are then entered via a keyboard. Entry to the room is granted if the digits keyed in coincide with four digits read from a magnetic track on the personal card. Three attempts at entry are permitted and if the correct number is not keyed in at the third attempt, an alarm sounds. Draw a flow chart to show the operation of this system.

2 Numbering systems

A MAIN POINTS CONCERNED WITH NUMBERING SYSTEMS

1 In everyday situations, a system of counting using a base of ten is employed. This is known as a **decimal** or **denary** system, and its main justification for use is often quoted as being that human beings have ten fingers/thumbs with which to count. The characteristics of a decimal numbering system are:
   (i) a set of ten distinct counting digits ($\emptyset$, 1, 2, 3, 4, 5, 6, 7, 8 & 9), and
   (ii) a place value (or weight) for each digit, organised in ascending powers of ten starting from the right.
   Thus, for example, the decimal number 2658_{10} may be considered as follows:

$$2658_{10} \text{ (decimal)} = 2 \times 1\emptyset^3 + 6 \times 1\emptyset^2 + 5 \times 1\emptyset^1 + 8 \times 1\emptyset^0$$
$$= 2\emptyset\emptyset\emptyset + 6\emptyset\emptyset + 5\emptyset + 8$$

2 A decimal system is not particularly suitable for direct use in electronic circuits. Due to practical limitations imposed by electronic devices, only two conditions are consistently predictable. These conditions are obtained when a chosen electronic device is made to act as a switch, and its two states are **on** and **off**, represented by the logic symbols 1 and $\emptyset$. This is known as 'two state logic', and each 1 or $\emptyset$ is called a bit (binary digit). A digital computer performs its tasks by manipulating information which is represented by patterns of bits.

3 One convenient method of representing numbers in terms of two-state logic is to make use of the **binary** (base two) system. The characteristics of a binary counting system are:
   (i) two counting digits (1 and $\emptyset$), and
   (ii) a place value (or weight) for each digit, organised in ascending powers of two starting from the right (see column 2 of *Table 1*).
   Thus, a 1 in a particular position of a binary number contributes its place value towards the total, but a $\emptyset$ contributes nothing. Therefore, the decimal equivalent of a binary number may be obtained by adding together all of the place values where a 1 occurs in that binary number. For example, consider the binary number $1\emptyset11\emptyset111_2$:

$$1\emptyset11\emptyset111_2 = 1 \times 2^7 + \emptyset \times 2^6 + 1 \times 2^5 + 1 \times 2^4 + \emptyset \times 2^3 + 1 \times 2^2 + 1 \times 2^1 + 1 \times 2^0$$
$$= 1 \times 128 + \emptyset \times 64 + 1 \times 32 + 1 \times 16 + \emptyset \times 8 + 1 \times 4 + 1 \times 2 + 1 \times 1$$
$$= 128 + \emptyset + 32 + 16 + \emptyset + 4 + 2 + 1$$
$$= 183_{10} \text{ (see } Problem\ 1).$$

Table 1 Comparison between decimal, binary, octal and hexadecimal numbering systems

Decimal	Binary	Octal	Hexadecimal
ØØ	ØØØØØØ	ØØ	ØØ
Ø1	ØØØØØ1	Ø1	Ø1
Ø2	ØØØØ1Ø	Ø2	Ø2
Ø3	ØØØØ11	Ø3	Ø3
Ø4	ØØØ1ØØ	Ø4	Ø4
Ø5	ØØØ1Ø1	Ø5	Ø5
Ø6	ØØØ11Ø	Ø6	Ø6
Ø7	ØØØ111	Ø7	Ø7
Ø8	ØØ1ØØØ	1Ø	Ø8
Ø9	ØØ1ØØ1	11	Ø9
1Ø	ØØ1Ø1Ø	12	ØA
11	ØØ1Ø11	13	ØB
12	ØØ11ØØ	14	ØC
13	ØØ11Ø1	15	ØD
14	ØØ111Ø	16	ØE
15	ØØ1111	17	ØF
16	Ø1ØØØØ	2Ø	1Ø
17	Ø1ØØØ1	21	11
18	Ø1ØØ1Ø	22	12
19	Ø1ØØ11	23	13
2Ø	Ø1Ø1ØØ	24	14
21	Ø1Ø1Ø1	25	15
22	Ø1Ø11Ø	26	16
23	Ø1Ø111	27	17
24	Ø11ØØØ	3Ø	18
25	Ø11ØØ1	31	19
26	Ø11Ø1Ø	32	1A
27	Ø11Ø11	33	1B
28	Ø111ØØ	34	1C

4   To carry out the reverse of the above process, that is, convert a decimal number into its equivalent binary number, several methods are possible. A system commonly adopted is the two's method. This method involves repeated division of a decimal number by 2 until a quotient of Ø is obtained. After each division, the remainder is noted (which is always 1 or Ø), and this forms the binary equivalent. For example, consider the decimal number 147_{10}. This may be converted into its binary equivalent as follows:

$$147 \div 2 = 73 \quad \text{remainder } 1$$
$$73 \div 2 = 36 \quad \text{remainder } 1$$
$$36 \div 2 = 18 \quad \text{remainder } Ø$$
$$18 \div 2 = 9 \quad \text{remainder } Ø$$
$$9 \div 2 = 4 \quad \text{remainder } 1$$
$$4 \div 2 = 2 \quad \text{remainder } Ø$$
$$2 \div 2 = 1 \quad \text{remainder } Ø$$
$$1 \div 2 = Ø \quad \text{remainder } 1$$

least significant digit (LSD) ... most significant digit (MSD)

Thus, the binary equivalent of 147_{10} is 1ØØ1ØØ11. (See *Problem 2*)

5 Although binary numbers are convenient for internal manipulation by a digital computer, they are tedious and error-prone when used by human operators. External to a computer, binary numbers may be represented by any convenient notation. For this purpose, however, a decimal system is not the most convenient, since it is not easy to relate each digit in a decimal number to specific groups of bits in its binary equivalent (this is because ten is not an exact power of two). **Octal** (base eight) or **hexadecimal** (base sixteen) representations are most frequently used since each group of three bits in a binary number does relate to a single octal digit, and each group of four bits in a binary number does relate to a single hexadecimal digit.

6 The characteristics of an **octal** numbering system are:
(i) eight distinct counting digits ($0, 1, 2, 3, 4, 5, 6$ and 7), and
(ii) a place value (or weight) for each digit, organised in ascending powers of eight starting from the right (see column 3 of *Table 1*).
Thus, for example, the octal number 352_8 may be considered as follows:

$$
\begin{aligned}
352_8 \text{ (octal)} &= 3 \times 8^2 + 5 \times 8^1 + 2 \times 8^0 \\
&= 3 \times 64 + 5 \times 8 + 2 \times 1 \\
&= 192 + 40 + 2 \\
&= 234_{10}. \text{ (See *Problem 3*)}
\end{aligned}
$$

7 In order to express a binary number in octal, its bits are arranged in groups of three, starting from the right, and an octal symbol is assigned to each group. For example, consider the binary number 101110010_2. This number may be expressed in its octal form by:
(i) grouping bits in threes from the right, and $\underset{5}{101} \quad \underset{6}{110} \quad \underset{2}{010}$
(ii) assigning octal symbols to each group.
Thus, $101110010_2 = 562_8$
(see *Problem 5*)

8 In order to convert an octal number into binary, the process in paragraph 7 is reversed. For example, consider the octal number 473_8. This number may be converted into binary form by:
(i) spacing out the octal digits, and $\underset{100}{4} \quad \underset{111}{7} \quad \underset{011}{3}$
(ii) converting each octal digit into binary form.
Thus, $473_8 = 100111011_2$. (See *Problem 6*)

9 The characteristics of a **hexadecimal** numbering system are:
(i) sixteen distinct counting digits ($0, 1, 2, 3, 4, 5, 6, 7, 8, 9, A, B, C, D, E$ and F), and
(ii) a place value (or weight) for each digit, organised in ascending powers of sixteen starting from the right (see column 4 of *Table 1*).
Since everyday situations make use of a decimal numbering system, only ten distinct counting digits have been devised. A hexadecimal system requires six more counting digits, and for this purpose, letters A to F are used. These letters correspond to decimal values 10 to 15. Thus, for example, the hexadecimal number $1A4E_{16}$ may be considered as follows:

$$
\begin{aligned}
1A4E_{16} \text{ (hex)} &= 1 \times 16^3 + A \times 16^2 + 4 \times 16^1 + E \times 16^0 \\
&= 1 \times 4096 + A \times 256 + 4 \times 16 + E \times 1 \\
&= 1 \times 4096 + 10 \times 256 + 4 \times 16 + 14 \times 1 \\
&= 4096 + 2560 + 64 + 14 \\
&= 6734_{10}. \text{ (See *Problem 7*)}
\end{aligned}
$$

10 In order to express a binary number in hexadecimal, its bits are arranged in groups of four, starting from the right, and a hexadecimal symbol is assigned to each group. For example, consider the binary number 1110011110101001_2. This number may be expressed in its hexadecimal form by:

(i) grouping bits in fours from the right, and $\underline{1110}$ $\underline{0111}$ $\underline{1010}$ $\underline{1001}$

(ii) assigning hexadecimal symbols to each group.   E   7   A   9

Thus, $1110011110101001_2 = E7A9_{16}$. (See *Problem 9*)

11 In order to convert a hexadecimal number into binary, the process in paragraph 10 is reversed. For example, consider the hexadecimal number $A3FB_{16}$. This number may be converted into binary form by:

(i) spacing out the hexadecimal digits, and   A   3   F   B

(ii) converting each hexadecimal digit into binary form.   1010   0011   1111   1011

Thus $A3FB_{16} = 1010001111111011_2$ (See *Problem 10*)

12 The rules for the **addition** of any two numbers of the same base are as follows:-

(i) the equation for addition is $X + Y = Z$, where X is called the **augend**, Y is called the **addend** and Z is called the **sum**;

(ii) digits in corresponding positions in each number are added, starting from the right; and

(iii) a **carry** into the next most significant position occurs if the sum of two digits equals or exceeds the value of the base used.

Due to the way in which a microprocessor operates (described in Chapter 3), it is necessary to differentiate between **carry out** and **carry in**. The difference between these two forms of carry may be studied by using the addition of 8_{10} and 9_{10} as an example.

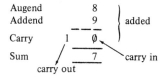

A **carry out** is generated because the sum of 8_{10} and 9_{10} exceeds the value of the base used (ten in this example). This carry out becomes a carry in for any subsequent addition, or, if no further addition takes place, the carry out must be considered as the most significant digit of the sum. A **carry in** to the least significant

Table 2

Augend	+	Addend	+	Carry in	=	Sum	,	Carry out
0		0		0		0		0
0		1		0		1		0
1		0		0		1		0
1		1		0		0		1
0		0		1		1		0
0		1		1		0		1
1		0		1		0		1
1		1		1		1		1

digit of an addition is not normally taken into account in ordinary arithmetic, but must be considered in most arithmetic operations when using a microprocessor.

13 When adding binary (base two) numbers, a carry out is generated when a sum equals or exceeds two. Binary addition rules are shown in *Table 2*. The application of the rules of binary addition may be studied by using the addition of $1110110\emptyset_2$ and $11111\emptyset1\emptyset_2$ as an example.

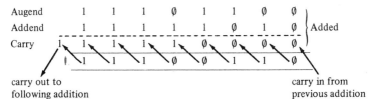

carry out to
following addition

carry in from
previous addition

In the absence of any following addition, the carry out becomes the most significant digit of the sum. **Therefore, $1110110\emptyset_2 + 11111\emptyset1\emptyset_2 = 11110\emptyset110_2$.**

Check:

$$1110110\emptyset_2 = 128 + 64 + 32 + 8 + 4 \qquad = 236_{10}$$
$$11111\emptyset1\emptyset_2 = 128 + 64 + 32 + 16 + 8 + 2 \quad = 25\emptyset_{10}$$
$$11110\emptyset110_2 = 256 + 128 + 64 + 32 + 4 + 2 \quad = 486_{10}$$

(See *Problem 11*)

14 Octal numbers may be added in a similar manner to that used for binary numbers, except that a carry is generated when a sum equals or exceeds eight. As an example of octal addition, consider the addition of 364_8 and 271_8.

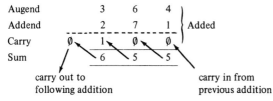

carry out to
following addition

carry in from
previous addition

Thus, $364_8 + 271_8 = 655_8$

Check:

$$364_8 = 3 \times 64 + 6 \times 8 + 4 \times 1 \quad = 244_{10}$$
$$271_8 = 2 \times 64 + 7 \times 8 + 1 \times 1 \quad = 185_{10}$$
$$655_8 = 6 \times 64 + 5 \times 8 + 5 \times 1 \quad = 429_{10}$$

(See *Problem 12*)

15 The procedure for adding hexadecimal numbers is similar to that used for octal numbers in paragraph 14, except that a carry is generated when a sum equals or exceeds sixteen. Initially, hexadecimal arithmetic operations may prove to be more difficult than arithmetic using the other bases so far discussed. This is because human operators are not familiar with the manipulation of numbers which

consist of a mixture of figures and letters. As an example of hexadecimal addition, consider the addition of $9A_{16}$ and $B7_{16}$.

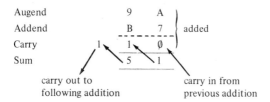

Augend	9	A
Addend	B	7
Carry	1 1	$\emptyset$
Sum	5	1

carry out to following addition

carry in from previous addition

Addition of the right hand column is as follows:

$$\begin{aligned} A_{16} + 7_{16} + \emptyset_{16} &= 1\emptyset_{10} + 7_{10} + \emptyset_{10} \\ &= 17_{10} \\ &= 11_{16} \text{ (i.e. sum 1, carry 1)} \end{aligned}$$

Addition of the next column is as follows:

$$\begin{aligned} 9_{16} + B_{16} + 1_{16} &= 9_{10} + 11_{10} + 1_{10} \\ &= 21_{10} \\ &= 15_{16} \text{ (i.e. sum 5, carry 1)} \end{aligned}$$

In the absence of any following addition, the carry out becomes the most significant digit of the sum. Therefore, $9A_{16} + B7_{16} = 151_{16}$.

Check:

$$\begin{aligned} 9A_{16} &= & 9 \times 16 + A \times 1 = 144_{10} + 10_{10} &= 154_{10} \\ B7_{16} &= & B \times 16 + 7 \times 1 = 176_{10} + 7_{10} &= 183_{10} \\ 151_{16} &= & 1 \times 256 + 5 \times 16 + 1 \times 1 = 256_{10} + 8\emptyset_{10} + 1_{10} &= 337_{10} \end{aligned}$$

(See *Problem 13*)

16 The rules for the **subtraction** of any two numbers of the same base are as follows:
   (i) the equation for subtraction is $X-Y = Z$, where X is called the **minuend**, Y is called the **subtrahend** and Z is called the **difference**;
   (ii) digits in the subtrahend are subtracted from digits in corresponding positions in the minuend, starting from the right;
   (iii) if the difference obtained in (ii) is less than zero (i.e. negative), a 'borrow in' from the next most significant position of the minuend is made. The value of a 'borrow in' is equal to the base used, and 1 (**borrow out**) must be added to the next most significant position of the subtrahend to compensate.

These rules may be studied by using decimal subtraction $42_{10}-19_{10}$ as an example.

Minuend	4	2 (+1$\emptyset$)		4	12
Subtrahend	1 (+1)	9	borrow	2	9
Difference	2	3		2	3

subtracted

Thus, $42_{10}-19_{10} = 23_{10}$

24

Table 3

Minuend	−	Subtrahend	=	Difference	Borrow
$\emptyset$		$\emptyset$		$\emptyset$	$\emptyset$
1		$\emptyset$		1	$\emptyset$
1		1		$\emptyset$	$\emptyset$
$\emptyset$		1		1	1

17 The rules of binary subtraction are summarised in *Table 3*. The application of the rules of binary subtraction may be studied by considering $1\emptyset1\emptyset\emptyset1_2 - 111\emptyset1\emptyset_2$ as an example.

Thus, $1\emptyset1\emptyset\emptyset1_2 - 111\emptyset1\emptyset_2 = 1\emptyset111_2$

$$\text{Check:} \ 1\emptyset1\emptyset\emptyset1_2 = 64 + 16 + 1 = 81_{10}$$
$$111\emptyset1\emptyset_2 = 32 + 16 + 8 + 2 = 58_{10}$$
$$1\emptyset111_2 = 16 + 4 + 2 + 1 = 23_{10} \ (\text{See } Problem \ 14)$$

18 Octal numbers may be subtracted in a similar manner to that used for binary numbers in paragraph 17, except that a 'borrow in' of eight occurs. For example, consider the octal subtraction $45_8 - 17_8$.

$$\text{Check:} \ 45_8 = 4 \times 8 + 5 \times 1 = 37_{10}$$
$$17_8 = 1 \times 8 + 7 \times 1 = 15_{10}$$
$$26_8 = 2 \times 8 + 6 \times 1 = 22_{10} \ (\text{See } Problem \ 15)$$

19 Hexadecimal numbers may be subtracted in a similar manner to that used for octal numbers in paragraph 18, except that a 'borrow in' of sixteen occurs. For example, consider the hexadecimal subtraction $A7_{16} - 6E_{16}$.

Subtraction of the right hand column is as follows:

$$17_{16} - E_{16} = 23_{10} - 14_{10} = 9_{10} = 9_{16}$$

Subtraction of the next column is as follows:

$$A_{16} - 7_{16} = 1\emptyset_{10} - 7_{10} = 3_{10} = 3_{16}$$

Thus, $A7_{16} - 6E_{16} = 39_{16}$

25

Check:

$$A7_{16} = A \times 16 + 7 \times 1 = 16\emptyset + 7 = 167_{10}$$
$$6E_{16} = 6 \times 16 + E \times 1 = 96 + 14 = \underline{11\emptyset_{10}}$$
$$39_{16} = 3 \times 16 + 9 \times 1 = 48 + 9 = \underline{57_{10}}$$

(See *Problem 16*)

20 In any arithmetic, the need frequently arises to represent **negative quantities**. This is done in ordinary arithmetic by using a minus sign, but this sign is not understood by a microprocessor. Therefore an alternative method must be found for representing negative quantities, in binary form, in a manner which is understood by a microprocessor. The usual method of doing this is to make the most significant bit (MSB) of a binary number act as a **sign bit** with the following meaning:

(i) for all positive values, the MSB is $\emptyset$, and

(ii) for all negative values, the MSB is 1.

This naturally reduces the magnitude of the largest number which may be represented by a given number of bits to half of its unsigned value. For example, an eight bit binary number may be treated as:

(i) **unsigned binary** with values in the range $\emptyset$ to 255_{10}, or

(ii) **signed binary** with values in the range $+127$ to -128_{10}.

Several different methods exist for the representation of negative binary numbers, the most common being:

(i) sign-magnitude; (ii) one's complement; and (iii) two's complement.

21 When using the **sign-magnitude** system, a negative number is formed by writing down its positive value and inverting the sign bit. Examples of sign-magnitude binary numbers are as follows:

		sign bit $\downarrow$	magnitude						
$+9_{10}$	$=$	$\emptyset$	$\emptyset$	$\emptyset$	$\emptyset$	1	$\emptyset$	$\emptyset$	1
-9_{10}	$=$	1	$\emptyset$	$\emptyset$	$\emptyset$	1	$\emptyset$	$\emptyset$	1
$+1\emptyset8_{10}$	$=$	$\emptyset$	1	1	$\emptyset$	1	1	$\emptyset$	$\emptyset$
$-1\emptyset8_{10}$	$=$	1	1	1	$\emptyset$	1	1	$\emptyset$	$\emptyset$

This system enables the value of a negative number to be readily calculated, but performing any arithmetic on numbers so represented is very difficult. For example, adding $+9_{10}$ to -9_{10} using the above values results in anything but the correct answer of zero.

22 A better method for representing negative numbers is to use a system of **complements**. A complement *completes* a number. For example, in a 9's complement system, the complement of a number is the value which must be added to that number to give a sum of 9. Thus, the 9's complement of 2 is 7 and the 9's complement of 4 is 5. Complements to almost any number may be formed, but in a binary system, only two types of complement are possible. These are:

(i) the one's complement, and (ii) the two's complement.

The principal advantage of using complements as far as a microprocessor is concerned is that subtraction may be performed by the addition of a number's complement. This simplifies the hardware of a microprocessor (see Chapter 3).

23 The **one's complement** of a number is formed by subtracting each digit of the number from one. For example, consider the binary number $000101 11_2$ (23_{10}):

1	1	1	1	1	1	1	1	} subtracted
0	0	0	1	0	1	1	1	
1	1	1	0	1	0	0	0	

Thus, the one's complement of 00010111_2 is 11101000_2.
It can be seen from this example that the one's complement of a binary number may be obtained by **inversion** of all its bits, that is, changing all of the 1's to 0's and all of the 0's to 1's. Therefore, using the one's complement notation, examples of negative numbers are are as follows:

$+9_{10} = 00001001_2$

$-9_{10} = 11110110_2$ (one's complement)

$+108_{10} = 01101100_2$

$-108_{10} = 10010011_2$ (one's complement)

One difficulty which arises from the use of one's complement notation is caused by the fact that the system has two values for zero:

$+0_{10} = 00000000_2$

$-0_{10} = 11111111_2$ (one's complement)

The effect of having two values for zero is to cause results to be one less than their true value when using one's complement notation.

24 The problem of having two values for zero may be overcome by representing negative numbers in **two's complement** form. The two's complement of a binary number is obtained by adding 1 to its one's complement. For example, consider -23_{10} in two's complement form:

$+23_{10} = 00010111_2$

$-23_{10} = 11101000$ (one's complement)   } added

$\phantom{-23_{10} = }00000001$

$\phantom{-23_{10} = }\overline{11101001}$ (two's complement)

Thus, in two's complement form, $-23_{10} = 11101001_2$. (See *Problems 17 and 18*)
The two's complement of a binary number may be obtained by writing down all bits from the least significant digit (LSD) up to, and including the first '1', in true form and complementing the remaining bits. Therefore, -56_{10} for example, may be written in two's complement form as follows:

$+56_{10} = 0\ 0\ 1\ 1\ 1\ 0\ 0\ 0_2$

first '1'

invert ⸺ true

$-56_{10} = 1\ 1\ 0\ 0\ 1\ 0\ 0\ 0$

Thus, in two's complement form, $-56_{10} = 11001000_2$

In general, the sum of a binary number and its two's complement is always zero.

25 As stated in paragraph 22, a microprocessor is best able to perform subtraction by adding complements. This method of subtraction may be illustrated by means of the following examples:

(a) *Subtraction of 00001100_2 from 01001110_2 using the one's complement method*

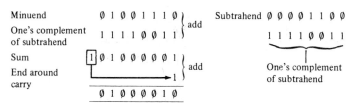

Since subtraction by the method of adding the one's complement gives a result one less than the true value, the answer must be corrected. This may be accomplished by adding the carry out from a sum to the sum. This is called '**end around carry**'.

Thus, from the above, $01001110_2 - 00001100_2 = 01000010_2$.

Check:
$$01001110_2 = 64 + 8 + 4 + 2 = 78_{10}$$
$$00001100_2 = 8 + 4 = 12_{10}$$
$$01000010_2 = 64 + 2 = 66_{10}$$

Note that if the subtrahend has fewer digits than the minuend (as in the above example), leading zeros must be added to the subtrahend before complementing it to give an identical number of digits in each.

(b) *Subtraction of 00110110_2 from 01001100_2 using the two's complement method*

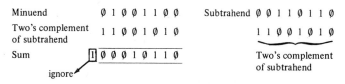

Thus, from the above, $01001100_2 - 00110110_2 = 00010110_2$

Check:
$$01001100_2 = 64 + 8 + 4 = 76_{10}$$
$$00110110_2 = 32 + 16 + 4 + 2 = 54_{10}$$
$$00010110_2 = 16 + 4 + 2 = 22_{10}$$

(See *Problem 20*)

26 Most printers and keyboards do not have the facility for dealing with suffixes to indicate the base of numbers used. Therefore the following prefixes are frequently used to indicate which base is being used:

(i) decimal – no prefix;

(ii) binary – %;

(iii) octal – @; and

(iv) hexadecimal – $ (or 'H' after the number).

Examples of the use of these symbols are as follows:

$$521 = 521_{10}$$
$$\%110010111 = 110010111_2$$
$$@245 = 245_8$$
$$\$6A = 6A_{16}$$
$$32H = 32_{16}$$

B   WORKED PROBLEMS ON NUMBERING SYSTEMS

Problem 1   Convert the following binary numbers into their decimal equivalents:
(a) 10111001_2; (b) 11100111_2; (c) 10000101_2; (d) 11001111_2.

(a) $10111001_2 = 1\times 2^7 + 0\times 2^6 + 1\times 2^5 + 1\times 2^4 + 1\times 2^3 + 0\times 2^2 + 0\times 2^1 + 1\times 2^0$
$= 1\times128 + 0\times 64 + 1\times 32 + 1\times 16 + 1\times 8 + 0\times 4 + 0\times 2 + 1\times 1$
$= 128 + 32 + 16 + 8 + 1 = 185_{10}$

(b) $11100111_2 = 1\times128 + 1\times64 + 1\times32 + 0\times16 + 0\times8 + 1\times4 + 1\times2 + 1\times1$
$= 128 + 64 + 32 + 4 + 2 + 1 = 231_{10}$

(c) $10000101_2 = 1\times128 + 0\times64 + 0\times32 + 0\times16 + 0\times8 + 1\times4 + 0\times2 + 1\times1$
$= 128 + 4 + 1 = 133_{10}$

(d) $11001111_2 = 1\times128 + 1\times64 + 0\times32 + 0\times16 + 1\times8 + 1\times4 + 1\times2 + 1\times1$
$= 128 + 64 + 8 + 4 + 2 + 1 = 207_{10}$

Problem 2   Convert the following decimal numbers into their binary equivalents by means of the two's method:
(a) 176_{10}; (b) 98_{10}; (c) 225_{10}; (d) 138_{10}.

(a) $176 \div 2 = 88$ remainder 0 LSD (least significant digit)
$88 \div 2 = 44$ remainder 0 ↑
$44 \div 2 = 22$ remainder 0
$22 \div 2 = 11$ remainder 0
$11 \div 2 = 5$ remainder 1
$5 \div 2 = 2$ remainder 1
$2 \div 2 = 1$ remainder 0
$1 \div 2 = 0$ remainder 1 MSD (most significant digit)

Thus, $176_{10} = 10110000_2$

(b) $98 \div 2 = 49$ remainder $\emptyset$ LSD
     $49 \div 2 = 24$ remainder 1   ↑
     $24 \div 2 = 12$ remainder $\emptyset$
     $12 \div 2 = 6$ remainder $\emptyset$
     $6 \div 2 = 3$ remainder $\emptyset$
     $3 \div 2 = 1$ remainder 1
     $1 \div 2 = \emptyset$ remainder 1   MSD
Thus, $98_{10} = 1100010_2$

(c) $225 \div 2 = 112$ remainder 1   LSD
     $112 \div 2 = 56$ remainder $\emptyset$ ↑
     $56 \div 2 = 28$ remainder $\emptyset$
     $28 \div 2 = 14$ remainder $\emptyset$
     $14 \div 2 = 7$ remainder $\emptyset$
     $7 \div 2 = 3$ remainder 1
     $3 \div 2 = 1$ remainder 1
     $1 \div 2 = \emptyset$ remainder 1   MSD
Thus, $225_{10} = 11100001_2$

(d) $138 \div 2 = 69$ remainder $\emptyset$ LSD
     $69 \div 2 = 34$ remainder 1   ↑
     $34 \div 2 = 17$ remainder $\emptyset$
     $17 \div 2 = 8$ remainder 1
     $8 \div 2 = 4$ remainder $\emptyset$
     $4 \div 2 = 2$ remainder $\emptyset$
     $2 \div 2 = 1$ remainder $\emptyset$
     $1 \div 2 = \emptyset$ remainder 1   MSD
Thus, $138_{10} = 10001010_2$

Problem 3 Convert the following octal numbers into their decimal equivalents:
(a) 325_8; (b) 427_8; (c) 173_8; (d) 714_8.

(a) $325_8 = 3 \times 8^2 + 2 \times 8^1 + 5 \times 8^0$
         $= 3 \times 64 + 2 \times 8 + 5 \times 1$
         $= 192 + 16 + 5 = 213_{10}$
Thus, $325_8 = 213_{10}$

(b) $427_8 = 4 \times 64 + 2 \times 8 + 7 \times 1$
         $= 256 + 16 + 7 = 279_{10}$
Thus, $427_8 = 279_{10}$

(c) $173_8 = 1 \times 64 + 7 \times 8 + 3 \times 1$
         $= 64 + 56 + 3 = 123_{10}$
Thus, $173_8 = 123_{10}$

(d) $714_8 = 7 \times 64 + 1 \times 8 + 4 \times 1$
         $= 448 + 8 + 4 = 460_{10}$
Thus, $714_8 = 460_{10}$

Problem 4 Convert the following decimal numbers into their octal equivalents:
(a) 129_{10}; (b) 400_{10}; (c) 53_{10}; (d) 199_{10}.

Decimal numbers may be converted into their octal equivalents by repeated division by eight, using a similar process to that employed in *Problem 2* for decimal to binary conversion.

(a) $129 \div 8 = 16$ remainder 1 $\blacktriangle$ LSD
     $16 \div 8 = 2$ remainder $\emptyset$
      $2 \div 8 = \emptyset$ remainder 2   MSD
Thus, $129_{10} = 2\emptyset1_8$

(b) $4\emptyset\emptyset \div 8 = 5\emptyset$ remainder $\emptyset$ $\blacktriangle$ LSD
     $5\emptyset \div 8 = 6$ remainder 2
      $6 \div 8 = \emptyset$ remainder 6   MSD
Thus, $4\emptyset\emptyset_{10} = 62\emptyset_8$

(c) $53 \div 8 = 6$ remainder 5 $\blacktriangle$ LSD
      $6 \div 8 = \emptyset$ remainder 6   MSD
Thus, $\emptyset53_{10} = \emptyset65_8$

(d) $199 \div 8 = 24$ remainder 7 $\blacktriangle$ LSD
     $24 \div 8 = 3$ remainder $\emptyset$
      $3 \div 8 = \emptyset$ remainder 3   MSD
Thus, $199_{10} = 3\emptyset7_8$

> *Problem 5*   Convert the following binary numbers into their octal equivalents:
> (a) $11\emptyset1\emptyset1111_2$; (b) $1\emptyset\emptyset\emptyset\emptyset111\emptyset_2$; (c) $1\emptyset1\emptyset1\emptyset\emptyset11_2$; (d) $111\emptyset111\emptyset_2$.

(a) grouping bits in threes from the right gives $\underbrace{11\emptyset}$ $\underbrace{1\emptyset1}$ $\underbrace{111}$

    assigning octal symbols to each group gives   6   5   7
Thus, $11\emptyset1\emptyset1111_2 = 657_8$

(b) grouping bits in threes from the right gives $\underbrace{1\emptyset\emptyset}$ $\underbrace{\emptyset\emptyset1}$ $\underbrace{11\emptyset}$

    assigning octal symbols to each group gives   4   1   6
Thus, $1\emptyset\emptyset\emptyset\emptyset111\emptyset_2 = 416_8$

(c) grouping bits in threes from the right gives $\underbrace{1\emptyset1}$ $\underbrace{\emptyset1\emptyset}$ $\underbrace{\emptyset11}$

    assigning octal symbols to each group gives   5   2   3
Thus, $1\emptyset1\emptyset1\emptyset\emptyset11_2 = 523_8$

(d) grouping bits in threes from the right gives $\underbrace{111}$ $\underbrace{\emptyset11}$ $\underbrace{1\emptyset1}$

    assigning octal symbols to each group gives   7   3   5
Thus, $111\emptyset111\emptyset1_2 = 735_8$

Problem 6 Convert the following octal numbers into their binary equivalents:
(a) 163_8; (b) 522_8; (c) 370_8; (d) 661_8.

(a) spacing out octal digits gives   1   6   3

     converting each octal digit into binary gives $\overbrace{0\,0\,1}$ $\overbrace{1\,1\,0}$ $\overbrace{0\,1\,1}$

Thus, $163_8 = 001110011_2$

(b) spacing out octal digits gives   5   2   2

     converting each octal digit into binary gives $\overbrace{1\,0\,1}$ $\overbrace{0\,1\,0}$ $\overbrace{0\,1\,0}$

Thus, $522_8 = 101010010_2$

(c) spacing out octal digits gives   3   7   0

     converting each octal digit into binary gives $\overbrace{0\,1\,1}$ $\overbrace{1\,1\,1}$ $\overbrace{0\,0\,0}$

Thus, $370_8 = 011111000_2$

(d) spacing out octal digits gives   6   6   1

     converting each octal digit into binary gives $\overbrace{1\,1\,0}$ $\overbrace{1\,1\,0}$ $\overbrace{0\,0\,1}$

Thus, $661_8 = 110110001_2$

Problem 7 Convert the following hexadecimal numbers into their decimal equivalents:
(a) $7A_{16}$; (b) $2F_{16}$; (c) $C9_{16}$; (d) BD_{16}.

(a) $7A_{16} = 7 \times 16 + 10 \times 1$ (from *Table 1*)
$= 112 \quad + 10$
$= 122_{10}$

Thus, $7A_{16} = 122_{10}$

(b) $2F_{16} = 2 \times 16 + 15 \times 1$
$= 32 \quad + 15$
$= 47_{10}$

Thus, $2F_{16} = 47_{10}$

(c) $C9_{16} = 12 \times 16 + 9 \times 1$
$= 192 \quad + 9$
$= 201_{10}$

Thus, $C9_{16} = 201_{10}$

(d) $BD_{16} = 11 \times 16 + 13 \times 1$
$= 176 \quad + 13$
$= 189_{10}$

Thus, $BD_{16} = 189_{10}$

Problem 8 Convert the following decimal numbers into their hexadecimal equivalents:
(a) 37_{10}; (b) 108_{10}; (c) 161_{10}; (d) 238_{10}.

Decimal numbers may be converted into their hexadecimal equivalents by repeated division by sixteen using a similar process to that employed in *Problem 4* for decimal to octal conversion.

(a) $37 \div 16 = 2$ remainder $5 \; (= 5_{16})$ ↑ LSD
    $2 \div 16 = \emptyset$ remainder $2 \; (= 2_{16})$ | MSD
Thus, $37_{10} = 25_{16}$

(b) $1\emptyset 8 \div 16 = 6$ remainder $12 \; (= C_{16})$ ↑ LSD
    $6 \div 16 = \emptyset$ remainder $6 \; (= 6_{16})$ | MSD
Thus, $1\emptyset 8_{10} = 6C_{16}$

(c) $161 \div 16 = 1\emptyset$ remainder $1 \; (= 1_{16})$ ↑ LSD
    $1\emptyset \div 16 = \emptyset$ remainder $1\emptyset \; (= A_{16})$ | MSD
Thus, $161_{10} = A1_{16}$

(d) $238 \div 16 = 14$ remainder $14 \; (= E_{16})$ ↑ LSD
    $14 \div 16 = \emptyset$ remainder $14 \; (= E_{16})$ | MSD
Thus, $238_{10} = EE_{16}$

Problem 9   Convert the following binary numbers into their hexadecimal equivalents:
(a) $11\emptyset 1\emptyset 111_2$; (b) $1\emptyset 1\emptyset \emptyset 11\emptyset_2$; (c) $11\emptyset \emptyset 1111_2$; (d) $1\emptyset \emptyset 11110_2$.

(a) grouping bits in fours from the right gives $1 1 \emptyset 1$ $\emptyset 1 1 1$

    assigning hexadecimal symbols to each group gives   D   7
    (see *Table 1*)
Thus, $11\emptyset 1\emptyset 111_2 = D7_{16}$

(b) grouping bits in fours from the right gives $1 \emptyset 1 \emptyset$ $\emptyset 1 1 \emptyset$

    assigning hexadecimal symbols to each group gives   A   6
Thus, $1\emptyset 1\emptyset \emptyset 11\emptyset_2 = A6_{16}$

(c) grouping bits in fours from the right gives $1 1 \emptyset \emptyset$ $1 1 1 1$

    assigning hexadecimal symbols to each group gives   C   F
Thus, $11\emptyset \emptyset 1111_2 = CF_{16}$

(d) grouping bits in fours from the right gives $1 \emptyset \emptyset 1$ $1 1 1 \emptyset$

    assigning hexadecimal symbols to each group gives   9   E
Thus, $1\emptyset \emptyset 11110_2 = 9E_{16}$

Problem 10   Convert the following hexadecimal numbers into their binary equivalents:
(a) $1F_{16}$; (b) $A5_{16}$; (c) $6A_{16}$; (d) $2B_{16}$.

(a) spacing out hexadecimal digits gives   1   F

converting each hexadecimal digit into binary gives $\overbrace{0\,0\,0\,1}$ $\overbrace{1\,1\,1\,1}$
(see *Table 1*)

Thus, $1F_{16} = 00011111_2$

(b) spacing out hexadecimal digits gives   A   5

converting each hexadecimal digit into binary gives $\overbrace{1\,0\,1\,0}$ $\overbrace{0\,1\,0\,1}$

Thus, $A5_{16} = 10100101_2$

(c) spacing out hexadecimal digits gives   6   A

converting each hexadecimal digit into binary gives $\overbrace{0\,1\,1\,0}$ $\overbrace{1\,0\,1\,0}$

Thus, $6A_{16} = 01101010_2$

(d) spacing out hexadecimal digits gives   2   B

converting each hexadecimal digit into binary gives $\overbrace{0\,0\,1\,0}$ $\overbrace{1\,0\,1\,1}$

Thus, $2B_{16} = 00101011_2$

Problem 11   Using the rules for binary addition, evaluate the following:
(a) $1011_2 + 1101_2$;   (c) $1010_2 + 1001_2$;
(b) $1100_2 + 1001_2$;   (d) $1111_2 + 100_2$.

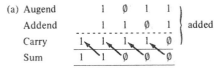

(a) Augend   1   0   1   1   ⎫
    Addend   1   1   0   1 ⎬ added
    Carry   1   1   1   1   0 ⎭
    Sum   1   1   0   0   0

Thus, $1011_2 + 1101_2 = 11000_2$

(b) Augend   1   1   0   0   ⎫
    Addend   1   0   0   1 ⎬ added
    Carry   1   0   0   0   0 ⎭
    Sum   1   0   1   0   1

Thus, $1100_2 + 1001_2 = 10101_2$

(c) Augend   1   0   1   0   ⎫
    Addend   1   0   0   1 ⎬ added
    Carry   1   0   0   0   0 ⎭
    Sum   1   0   0   1   1

Thus, $1010_2 + 1001_2 = 10011_2$

(d) Augend   1   1   1   1
    Addend $\emptyset$ 1 $\emptyset$ $\emptyset$ $\Big\}$ added
    Carry   1   1 $\emptyset$ $\emptyset$ $\emptyset$
    Sum   1 $\emptyset$ $\emptyset$ 1   1

Thus, $1111_2 + 1\emptyset\emptyset_2 = 1\emptyset\emptyset11_2$

Problem 12   Using the rules for octal addition, evaluate the following:
(a) $22_8 + 17_8$; (b) $37_8 + 25_8$; (c) $77_8 + 16_8$; (d) $65_8 + 43_8$.

(a) Augend   2   2
    Addend   1   7 $\Big\}$ added
    Carry   1 $\emptyset$
    Sum   4   1

Thus, $22_8 + 17_8 = 41_8$

(b) Augend   3   7
    Addend   2   5 $\Big\}$ added
    Carry   1 $\emptyset$
    Sum   6   4

Thus, $37_8 + 25_8 = 64_8$

(c) Augend   7   7
    Addend   1   6 $\Big\}$ added
    Carry   1   1 $\emptyset$
    Sum   1   1   5

Thus, $77_8 + 16_8 = 115_8$

(d) Augend   6   5
    Addend   4   3 $\Big\}$ added
    Carry   1   1 $\emptyset$
    Sum   1   3 $\emptyset$

Thus, $65_8 + 43_8 = 13\emptyset_8$

Problem 13   Using the rules for hexadecimal addition, evaluate the following:
(a) $47_{16} + 36_{16}$; (b) $3B_{16} + 81_{16}$; (c) $89_{16} + A6_{16}$; (d) $9A_{16} + BD_{16}$.

(a) Augend   4   7
    Addend   3   6 $\Big\}$ added
    Carry $\emptyset$ $\emptyset$
    Sum   7   D

Thus, $47_{16} + 36_{16} = 7D_{16}$

(b) Augend   3   B
    Addend   8   1 $\Big\}$ added
    Carry $\emptyset$ $\emptyset$
    Sum   B   C

Thus, $3B_{16} + 81_{16} = BC_{16}$

(c) Augend   8   9
    Addend   A   6 $\Big\}$ added
    Carry   1 $\emptyset$ $\emptyset$
    Sum   1   2   F

Thus, $89_{16} + A6_{16} = 12F_{16}$

(d) Augend   9   A
    Addend   B   D $\Big\}$ added
    Carry   1   1 $\emptyset$
    Sum   1   5   7

Thus, $9A_{16} + BD_{16} = 157_{16}$

Problem 14   Using the rules for binary subtraction, evaluate the following:
(a) $11\emptyset\emptyset_2 - 1\emptyset\emptyset1_2$; (b) $1\emptyset\emptyset1_2 - 1\emptyset1_2$; (c) $111\emptyset_2 - 1\emptyset11_2$; (d) $1\emptyset\emptyset_2 - 11_2$.

(a) Minuend   1   1 $\emptyset$(+2) $\emptyset$(+2) ⎫
    Subtrahend   1 $\emptyset$(+1) $\emptyset$(+1)   1   ⎬ subtracted
    Difference $\emptyset$ $\emptyset$ 1   1   ⎭

Thus, $1100_2 - 1001_2 = 11_2$

(b) Minuend   1 $\emptyset$(+2) $\emptyset$ 1   ⎫
    Subtrahend $\emptyset$(+1)   1 $\emptyset$ 1   ⎬ subtracted
    Difference $\emptyset$ 1 $\emptyset$ $\emptyset$ ⎭

Thus, $1001_2 - 101_2 = 100_2$

(c) Minuend   1   1   1(+2) $\emptyset$(+2) ⎫
    Subtrahend   1 $\emptyset$(+1)   1(+1)   1   ⎬ subtracted
    Difference $\emptyset$ $\emptyset$ 1   1   ⎭

Thus, $1110_2 - 1011_2 = 11_2$

(d) Minuend   1 $\emptyset$(+2) $\emptyset$(+2)   1   ⎫
    Subtrahend $\emptyset$(+1) $\emptyset$(+1)   1   1   ⎬ subtracted
    Difference $\emptyset$ 1   1 $\emptyset$ ⎭

Thus, $1001_2 - 11_2 = 110_2$

Problem 15   Using the rules for octal subtraction, evaluate the following:
(a) $37_8 - 12_8$; (b) $62_8 - 15_8$; (c) $51_8 - 36_8$; (d) $72_8 - 17_8$.

(a) Minuend   3   7   ⎫
    Subtrahend   1   2   ⎬ subtracted
    Difference   2   5

Thus, $37_8 - 12_8 = 25_8$

(b) Minuend   6   2(+8) ⎫
    Subtrahend   1(+1)   5   ⎬ subtracted
    Difference   4   5

Thus, $62_8 - 15_8 = 45_8$

(c) Minuend   5   1(+8) ⎫
    Subtrahend   3(+1)   6   ⎬ subtracted
    Difference   1   3

Thus, $51_8 - 36_8 = 13_8$

(d) Minuend   7   2(+8) ⎫
    Subtrahend   1(+1)   7   ⎬ subtracted
    Difference   5   3

Thus, $72_8 - 17_8 = 53_8$

Problem 16 Using the rules for hexadecimal subtraction, evaluate the following:
(a) $3C_{16} - 2A_{16}$; (b) $91_{16} - 2C_{16}$; (c) $60_{16} - 08_{16}$; (d) $FF_{16} - A9_{16}$.

(a) Minuend   3   C   } subtracted
     Subtrahend   2   A
     Difference   1   2   Thus, $3C_{16} - 2A_{16} = 12_{16}$

(b) Minuend   9   1(+16) } subtracted
     Subtrahend   2(+1)   C
     Difference   6   5   Thus, $91_{16} - 2C_{16} = 65_{16}$

(c) Minuend   6   0(+16) } subtracted
     Subtrahend   0(+1)   8
     Difference   5   8   Thus, $60_{16} - 08_{16} = 58_{16}$

(d) Minuend   F   F   } subtracted
     Subtrahend   A   9
     Difference   5   6   Thus, $FF_{16} - A9_{16} = 56_{16}$

Problem 17 Determine the two's complement of the following binary numbers:
(a) 10011100_2; (b) 01101011_2; (c) 11011010_2; (d) 10100110_2.

(a) Binary number   1 0 0 1 1 1 0 0 ——┐
                                  inverted

One's complement   0 1 1 0 0 0 1 1 } ←┘
                               1 } added

Two's complement   0 1 1 0 0 1 0 0

Therefore, **the two's complement of 10011100_2 is 01100100_2**

(b) Binary number   0 1 1 0 1 0 1 1 ——┐
                                    inverted

One's complement   1 0 0 1 0 1 0 0 } ←┘
                               1 } added

Two's complement   1 0 0 1 0 1 0 1

Therefore, **the two's complement of 01101011_2 is 10010101_2**

(c) Binary number   1 1 0 1 1 0 1 0 ——┐
                                    inverted

One's complement   0 0 1 0 0 1 0 1 } ←┘
                               1 } added

Two's complement   0 0 1 0 0 1 1 0

Therefore, **the two's complement of 11011010_2 is 00100110_2**

37

(d) Binary number   1 $\emptyset$ 1 $\emptyset$ $\emptyset$ 1 1 $\emptyset$ ⸺┐

                                            inverted

One's complement $\emptyset$ 1 $\emptyset$ 1 1 $\emptyset$ $\emptyset$ 1 } ◄⸺┘

                                            1 } added

Two's complement $\emptyset$ 1 $\emptyset$ 1 1 $\emptyset$ 1 $\emptyset$

Therefore, **the two's complement of $1\emptyset1\emptyset\emptyset11\emptyset_2$ is $\emptyset1\emptyset11\emptyset1\emptyset_2$**

Problem 18   Show how the following decimal numbers may be represented by using eight bit two's complement notation:
(a) -35_{10}; (b) $-1\emptyset5_{10}$; (c) -81_{10}; (d) -28_{10}.

(a) Using the two's method, $+35_{10} = \emptyset\ \emptyset\ 1\ \emptyset\ \emptyset\ \emptyset\ 1\ 1_2$ ⸺┐

                                          inverted

One's complement   1 1 $\emptyset$ 1 1 1 $\emptyset$ $\emptyset$ } ◄⸺┘

                                          1 } added

Two's complement   1 1 $\emptyset$ 1 1 1 $\emptyset$ 1

Therefore, **in two's complement notation**, $-35_{10} = 11\emptyset1111\emptyset1_2$

(b) Using the two's method, $+1\emptyset5_{10} = \emptyset\ 1\ 1\ \emptyset\ 1\ \emptyset\ \emptyset\ 1_2$ ⸺┐

                                          inverted

One's complement   1 $\emptyset$ $\emptyset$ 1 $\emptyset$ 1 1 $\emptyset$ } ◄⸺┘

                                          1 } added

Two's complement   1 $\emptyset$ $\emptyset$ 1 $\emptyset$ 1 1 1

Therefore, **in two's complement notation**, $-1\emptyset5_{10} = 1\emptyset\emptyset1\emptyset111_2$

(c) Using the two's method, $+81_{10} = \emptyset\ 1\ \emptyset\ 1\ \emptyset\ \emptyset\ \emptyset\ 1_2$ ⸺┐

                                          inverted

One's complement   1 $\emptyset$ 1 $\emptyset$ 1 1 1 $\emptyset$ } ◄⸺┘

                                          1 } added

Two's complement   1 $\emptyset$ 1 $\emptyset$ 1 1 1 1

Therefore, **in two's complement notation**, $-81_{10} = 1\emptyset1\emptyset1111_2$

(d) Using the two's method, $+28_{10} = \emptyset\ \emptyset\ \emptyset\ 1\ 1\ 1\ \emptyset\ \emptyset$ ⸺┐

                                          inverted

One's complement   1 1 1 $\emptyset$ $\emptyset$ $\emptyset$ 1 1 } ◄⸺┘

                                          1 } added

Two's complement   1 1 1 $\emptyset$ $\emptyset$ 1 $\emptyset$ $\emptyset$

Therefore, **in two's complement notation**, $-28_{10} = 111\emptyset\emptyset1\emptyset\emptyset_2$

Problem 19   Convert the following decimal numbers into their equivalent hexadecimal values:
(a) -6_{10}; (b) $-5\emptyset_{10}$; (c) -95_{10}; (d) -22_{10}.

There are several methods by which this problem may be solved, such as:

(i) determining the equivalent two's complement value (as in *Problem 18*) and converting into hexadecimal; or

(ii) evaluating the magnitude of the number as a positive hexadecimal value (as in *Problem 8*) and subtract this value from zero by application of the rules for hexadecimal subtraction.

Note that leading 'F's in negative hexadecimal numbers are similar to leading '∅'s in positive numbers and those outside of the normal operating field may be ignored.

Using method (ii):

(a) $-6_{10} = \emptyset\emptyset_{10} - (+\emptyset6_{10}) = \emptyset\emptyset_{16} - (+\emptyset6_{16})$
  using decimal to hexadecimal conversion (see *Problem 8*)

		$\emptyset(+16)$	$\emptyset(+16)$	
Minuend				} subtracted
Subtrahend	$\emptyset(+1)$	$\emptyset(+1)$	6	
Difference		F	A	

Thus, $-6_{10} = FA_{16}$

(b) $-5\emptyset_{10} = \emptyset\emptyset_{10} - (+5\emptyset_{10}) = \emptyset\emptyset_{16} - (+32_{16})$
  using decimal to hexadecimal conversion (see *Problem 8*)

		$\emptyset(+16)$	$\emptyset(+16)$	
Minuend				} subtracted
Subtrahend	$\emptyset(+1)$	$3(+1)$	2	
Difference		C	E	

Thus, $-5\emptyset_{10} = CE_{16}$

(c) $-95_{10} = \emptyset\emptyset_{10} - (+95_{10}) = \emptyset\emptyset_{16} - (+5F_{16})$
  using decimal to hexadecimal conversion (see *Problem 8*)

		$\emptyset(+16)$	$\emptyset(+16)$	
Minuend				} subtracted
Subtrahend	$\emptyset(+1)$	$5(+1)$	F	
Difference		A	1	

Thus, $-95_{10} = A1_{16}$

(d) $-22_{10} = \emptyset\emptyset_{10} - (+22_{10}) = \emptyset\emptyset_{16} - (+16_{16})$
  using decimal to hexadecimal conversion (see *Problem 8*)

		$\emptyset(+16)$	$\emptyset(+16)$	
Minuend				} subtracted
Subtrahend	$\emptyset(+1)$	$1(+1)$	6	
Difference		E	A	

Thus, $-22_{10} = EA_{16}$

Problem 20 Evaluate the following by using the two's complement method for subtraction:

(a) $10110101_2 - 101111_2$; (c) $10001011_2 - 1011101_2$;

(b) $11101110_2 - 1101_2$; (d) $11111011_2 - 10010110_2$.

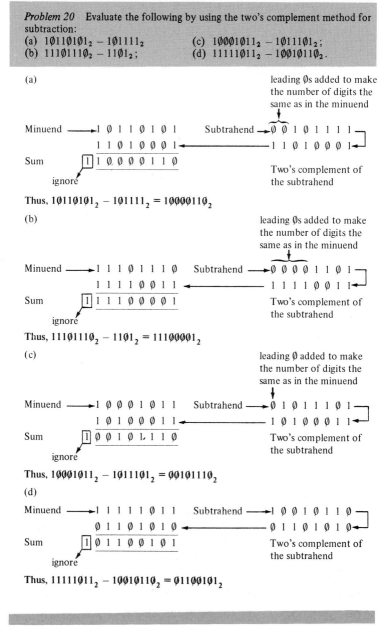

(a)

leading 0s added to make the number of digits the same as in the minuend

Minuend ⟶ 1 0 1 1 0 1 0 1   Subtrahend ⟶ 0 0 1 0 1 1 1 1 ⟶

  1 1 0 1 0 0 0 1 ⟵   1 1 0 1 0 0 0 1 ⟵

Sum   [1] 1 0 0 0 0 1 1 0

ignore

Two's complement of the subtrahend

Thus, $10110101_2 - 101111_2 = 10000110_2$

(b)

leading 0s added to make the number of digits the same as in the minuend

Minuend ⟶ 1 1 1 0 1 1 1 0   Subtrahend ⟶ 0 0 0 0 1 1 0 1 ⟶

  1 1 1 1 0 0 1 1 ⟵   1 1 1 1 0 0 1 1 ⟵

Sum   [1] 1 1 1 0 0 0 0 1

ignore

Two's complement of the subtrahend

Thus, $11101110_2 - 1101_2 = 11100001_2$

(c)

leading 0 added to make the number of digits the same as in the minuend

Minuend ⟶ 1 0 0 0 1 0 1 1   Subtrahend ⟶ 0 1 0 1 1 1 0 1 ⟶

  1 0 1 0 0 0 1 1 ⟵   1 0 1 0 0 0 1 1 ⟵

Sum   [1] 0 0 1 0 1 1 1 0

ignore

Two's complement of the subtrahend

Thus, $10001011_2 - 1011101_2 = 00101110_2$

(d)

Minuend ⟶ 1 1 1 1 1 0 1 1   Subtrahend ⟶ 1 0 0 1 0 1 1 0 ⟶

  0 1 1 0 1 0 1 0 ⟵   0 1 1 0 1 0 1 0 ⟵

Sum   [1] 0 1 1 0 0 1 0 1

ignore

Two's complement of the subtrahend

Thus, $11111011_2 - 10010110_2 = 01100101_2$

40

C FURTHER PROBLEMS ON NUMBERING SYSTEMS

(a) MULTI-CHOICE PROBLEMS (answers on page 266)

In Problems 1 to 10 select the correct answer from those given.

1 The one's complement of $1011011 0_2$ is:
(a) 10110110_2; (c) 01101101_2;
(b) 01001001_2; (d) 01011011_2.

2 The two's complement of 11010011_2 is:
(a) 11010011_2; (c) 00101101_2;
(b) 00101100_2; (d) 10100110_2.

3 The maximum range of values that may be represented by an eight bit signed (two's complement) number are:
(a) -127_{10} to $+128_{10}$; (c) -255_{10} to $+256_{10}$;
(b) -256_{10} to $+255_{10}$; (d) -128_{10} to $+127_{10}$.

4 In two's complement notation, the binary number 10111001_2 has a value of:
(a) -71_{10}; (b) $+71_{10}$; (c) -185_{10}; (d) $+185_{10}$.

5 Electronic circuits in microprocessors and digital computers process:
(a) decimal numbers; (b) octal numbers; (c) hexadecimal numbers; (d) binary numbers.

6 In order to represent decimal numbers in the range $\emptyset$ to $+65\emptyset\emptyset\emptyset_{10}$, a hexadecimal number must have:
(a) 5 digits; (b) 3 digits; (c) 4 digits; (d) 6 digits.

7 The result of adding 26_{16} to 37_{16} is:
(a) 63_{16}; (b) $6D_{16}$; (c) 65_{16}; (d) $5D_{16}$.

8 Which of the following is correct?
(a) $\emptyset7_{16} = \emptyset7_{10}$; (b) $17_{10} = 17_8$; (c) $\emptyset7_{10} > \emptyset7_8$; (d) $\emptyset7_{16} > \emptyset7_{10}$.

9 If $8\emptyset_{16}$ is converted into two's complement binary form:
(a) the MSD is 1 and signifies a negative number;
(b) the MSD is $\emptyset$ and signifies a positive number;
(c) the MSD is 1 and signifies a positive number;
(d) the MSD is $\emptyset$ and signifies a negative number.

10 The result of subtracting $\emptyset9_{10}$ from 32_{16} is:
(a) 13_{10}; (b) 29_{16}; (c) 13_{16}; (d) 19_{10}.

(b) CONVENTIONAL PROBLEMS

1 Convert the following binary numbers into their (unsigned) decimal equivalents:
(a) 11110001_2; $[241_{10}]$ (c) 11001100_2; $[204_{10}]$
(b) 10101010_2; $[170_{10}]$ (d) 10001101_2. $[141_{10}]$

2 Convert the following decimal numbers into binary form by using the two's method:
(a) 240_{10}; $[11110000_2]$ (c) 163_{10} $[10100011_2]$
(b) 87_{10}; $[1010111_2]$ (d) 149_{10}. $[10010101_2]$

3 Convert the following octal numbers into their decimal equivalents:
   (a) 225_8;   [149_{10}]   (c) 745_8 [485_{10}]
   (b) $1Ø1_8$;   [65_{10}]   (d) 176_8.   [$19Ø_{10}$]

4 Convert the following decimal numbers into their octal equivalents:
   (a) 284_{10};   [434_8]   (c) $1ØØ_{10}$;   [144_8]
   (b) 659_{10};   [1223_8]   (d) 325_{10}.   [$5Ø5_8$]

5 Convert the following binary numbers into their octal equivalents:
   (a) $11ØØ1Ø111_2$;   [627_8]   (c) $1ØØØ11Ø1Ø_2$;   [432_8]
   (b) $111Ø111Ø_2$;   [356_8]   (d) $1Ø111Ø111_2$.   [567_8]

6 Convert the following octal numbers into their binary equivalents:
   (a) 361_8;   [$11110ØØ1_2$]   (c) 122_8;   [$1Ø1ØØ1Ø_2$]
   (b) 746_8;   [111100110_2] (d) 37_8.   [11111_2]

7 Convert the following hexadecimal numbers into their (unsigned) decimal equivalents:
   (a) $E5_{16}$;   [229_{10}]   (c) 98_{16};   [152_{10}]
   (b) $2C_{16}$;   [44_{10}]   (d) $F1_{16}$.   [241_{10}]

8 Convert the following decimal values into their hexadecimal equivalents:
   (a) 54_{10};   [36_{16}]   (c) 91_{10};   [$5B_{16}$]
   (b) $2ØØ_{10}$;   [$C8_{16}$]   (d) 238_{10}.   [EE_{16}]

9 Convert the following binary numbers into their hexadecimal equivalents:
   (a) $11Ø1Ø111_2$;   [$D7_{16}$]   (c) $1ØØØ1Ø11_2$;   [$8B_{16}$]
   (b) $111Ø1Ø1Ø_2$;   [EA_{16}]   (d) $1Ø1ØØ1Ø1_2$.   [$A5_{16}$]

10 Convert the following hexadecimal numbers into their binary equivalents:
   (a) 37_{16};   [$11Ø1Ø111_2$]   (c) $9F_{16}$;   [$1ØØ11111_2$]
   (b) ED_{16};   [$111Ø11Ø1_2$]   (d) $A2_{16}$.   [$1Ø1ØØØ1Ø_2$]

11 Using the rules for binary addition, evaluate the following:
   (a) $1Ø1Ø_2 + 111_2$;   [$1ØØØ1_2$]   (c) $1Ø11_2 + 1Ø1Ø_2$;   [$1Ø1Ø1_2$]
   (b) $1111_2 + 1ØØ1_2$;   [$11ØØØ_2$]   (d) $111Ø_2 + 11Ø_2$.   [$1Ø1ØØ_2$]

12 Using the rules for octal addition, evaluate the following:
   (a) $33_8 + 27_8$;   [62_8]   (c) $12_8 + 7_8$;   [21_8]
   (b) $76_8 + 54_8$;   [152_8]   (d) $67_8 + 17_8$.   [$1Ø6_8$]

13 Using the rules for hexadecimal addition, evaluate the following:
   (a) $3E_{16} + 2C_{16}$;   [$6A_{16}$]   (c) $6F_{16} + 29_{16}$;   [98_{16}]
   (b) $A8_{16} + 9B_{16}$;   [143_{16}]   (d) $BB_{16} + 1A_{16}$.   [$D5_{16}$]

14 Using the rules for binary subtraction, evaluate the following:
   (a) $1111_2 - 1Ø1_2$;   [$1Ø1Ø_2$]   (c) $111Ø_2 - 1Ø1_2$;   [$1ØØ1_2$]
   (b) $1Ø11_2 - 1ØØ_2$;   [111_2]   (d) $1Ø1Ø_2 - 1ØØ1_2$.   [1_2]

15 Using the rules for octal subtraction, evaluate the following:
   (a) $62_8 - 35_8$;   [25_8]   (c) $24_8 - 15_8$;   [7_8]
   (b) $71_8 - 16_8$;   [53_8]   (d) $76_8 - Ø7_8$.   [67_8]

16 Using the rules for hexadecimal subtraction, evaluate the following:
   (a) $A5_{16} - 68_{16}$;   [$3D_{16}$]   (c) $9F_{16} - 3D_{16}$;   [62_{16}]
   (b) $F2_{16} - BC_{16}$;   [36_{16}]   (d) $81_{16} - 1B_{16}$.   [66_{16}]

17 Determine the two's complement of the following binary numbers:
   (a) 10011100_2; $[01100100_2]$ (c) 10101010_2; $[01010110_2]$
   (b) 11110000_2; $[00010000_2]$ (d) 11011100_2. $[00100100_2]$

18 Show how the following decimal numbers may be represented by using eight bit two's complement notation:-
   (a) -120_{10}; $[10001000_2]$ (c) -65_{10}; $[10111111_2]$
   (b) -2_{10}; $[11111110_2]$ (d) -93_{10}. $[10100011_2]$

19 Convert the following decimal numbers into their equivalent hexadecimal values:
   (a) -20_{10}; $[EC_{16}]$ (c) -36_{10}; $[DC_{16}]$
   (b) -115_{10}; $[8D_{16}]$
            (d) -99_{10} $[9D_{16}]$

20 Evaluate the following by using the two's complement method for subtraction:
   (a) $11110111_2 - 11011_2$;   (c) $10110110_2 - 10000001_2$;
            $[11011100_2]$ $[110101_2]$
   (b) $10000000_2 - 11111_2$;
            $[1100001_2]$ (d) $11010010_2 - 10000111_2$.

                                             $[1001011_2]$

3 Microprocessor-based systems

A MAIN POINTS CONCERNED WITH THE OPERATION OF MICROPROCESSOR-BASED SYSTEMS

1   It is necessary to provide storage for binary data within a microprocessor whilst manipulation of that data takes place. The basic storage element for a binary digit (bit) is a **bistable** or **flip-flop** circuit (see *Problem 1*). Data are usually manipulated by a microprocessor in the form of groups of eight or sixteen bits. Therefore, groups of eight or sixteen bistable or flip-flop circuits are required for the temporary storage of data within a microprocessor (whilst manipulation takes place). Such a group of circuits which handle groups of related bits is called a **register** (see *Figure 1*).

2   A **microprocessor** or **MPU** (microprocessing unit) performs the function of a CPU (central processing unit), and therefore contains many registers. The exact number and types of registers used varies from one microprocessor design to another.
A simplified microprocessor block diagram is illustrated in *Figure 2*. The function of each of the circuit blocks of *Figure 2* may be summarised as follows:

(i) **Instruction register (IR).** This register is used to store an instruction, thereby enabling a microprocessor to decode the instruction and determine what action to take. The instruction contained within this register is the **current instruction**, that is, the instruction being (or about to be) executed.

(ii) **Program counter (PC).** The instructions which form a microprocessor program must be executed in **strict numerical sequence**. The program counter is a sixteen bit register which is capable of being incremented by a count of one after each instruction has been carried out and therefore acts as a sixteen bit binary counter. The program counter is, therefore, responsible for generating the information which is put out on the address bus and ensures that instructions are carried out in

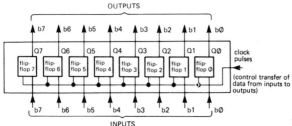

Figure 1

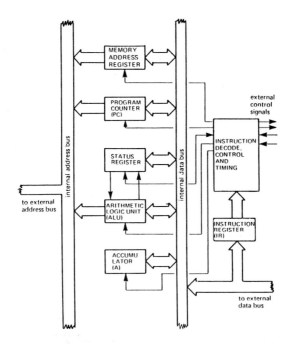

Figure 2

the correct sequence. While a microprocessor is carrying out an instruction, its program counter usually contains the address of the **next instruction** to be executed.

(iii) **Memory address register** (MAR). This is a sixteen bit register which is loaded with the address of a single memory location. Certain types of microprocessor instructions specify data which is to be manipulated as that pointed to by one of the store address registers. For this reason, a store address register is frequently called a **'memory pointer register'**. The exact type and number of store address registers varies from one microprocessor to another.

(iv) **Accumulator (A)**. This register is used to store one of the inputs to the **arithmetic-logic unit** (ALU), and as such is one of the most-used registers in a

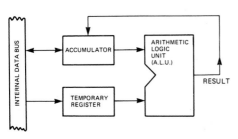

microprocessor. In fact, most microprocessor instructions involve use of the accumulator in one form or another. The results of all arithmetical and logical operations are saved in the accumulator, since the ALU output is connected back into the accumulator (see *Figure 3*). Results are therefore accumulated in this register — hence the term 'accumulator'.

Figure 3

45

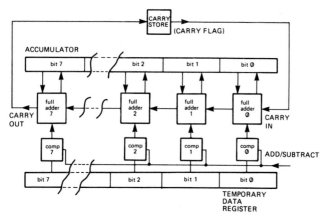

Figure 4

(v) **Arithmetic-logic unit (ALU).** The arithmetic-logic unit is responsible for carrying out arithmetical and logical operations involving its two inputs. The following operations are available on the arithmetic-logic units of most microprocessors:

(a) add
(b) subtract } arithmetical operations

(c) OR
(d) AND } logical operations
(e) EXCLUSIVE-OR

An arithmetic-logic unit consists of a group of eight (or sixteen) circuit elements called '**full adders**' (see *Problem 2*), arranged as shown in *Figure 4*. **Complementers** are used to enable subtraction to be performed.

(vi) **Status register.** This register is also known as a '**flag register**' or '**condition code register**', and contains information concerning the result of the latest process carried out in the arithmetic-logic unit. A status register is not, strictly speaking,

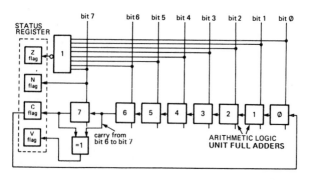

Figure 5

46

Table 1

FLAG	*SET* (= 1)	*RESET* (= ∅)
Z	result zero	result not zero
N	result negative	result positive
C	carry generated	carry not generated
V	overflow occurs	overflow does not occur

a register, but a collection of unrelated bistables or flip-flops which for convenience are grouped together and considered as a register. Each bistable or flip-flop in a status register is frequently called a '**flag**', and may be used to indicate that the following conditions are detected:

(a) a result is zero; (c) a carry out occurs; and
(b) a result is negative; (d) an overflow occurs.

(See *Problems 3, 4 and 5*)

Interpretation of the condition shown by each flag may be obtained by referring to *Table 1*. A simplified diagram which shows how flags are connected to an

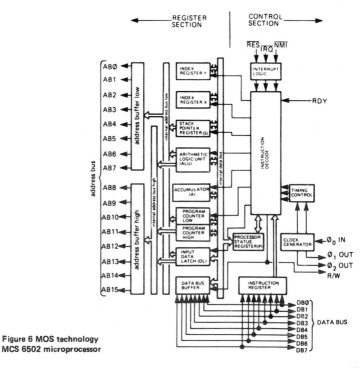

Figure 6 MOS technology
MCS 6502 microprocessor

47

arithmetic-logic unit is depicted in *Figure 5* (see *Appendix A* for information regarding logic gates).

(vii) **Control and timing devices.** This section of a microprocessor is responsible for organisation of data flow between the various circuits in response to an instruction, and for the timing of such data transfers. To enable this to take place, this section of a microprocessor must be able to determine the exact nature of an instruction (**instruction decode**), and control all of the logic circuits through which data passes (see *Problem 7*). An additional function performed by this section is that of generating such **control signals** as are required by circuits external to a microprocessor, for example, the **read/write control** signal for a system random access memory (RAM).

Block diagrams for three commonly used microprocessors are illustrated in *Figures 6, 7 and 8.*

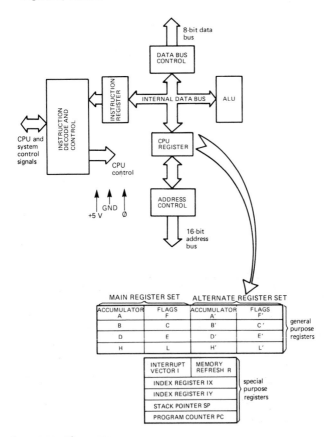

Figure 7 Zilog/Mostek Z80 microprocessor

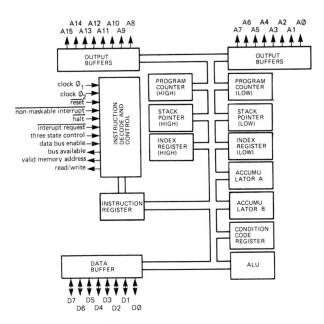

Figure 8 Motorola MC 6800 microprocessor

3 The microprocessor alone is capable of doing very little. It requires a certain amount of supporting hardware, which in a minimal system consists of **memory** (ROM and RAM), and **input/output** (I/O) circuits. These are connected together as shown in *Figure 9*, and form what is called **microcomputer**. The function of each block is as follows:

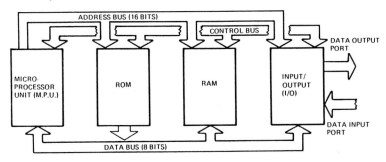

Figure 9 Microcomputer block diagram

49

(a) The **central processing unit** (CPU) is another name for the microprocessor. It is responsible for carrying out the following processes:
   (i) Fetching instructions from the memory in the correct sequence.
   (ii) Interpreting each instruction as it is received.
   (iii) Acting upon each received instruction.
   (iv) Generating the necessary control signals to accomplish (i) to (iii).
   The instructions that the microprocessor is able to act upon include a wide variety of arithmetic, logic and data transfer functions.

(b) The **memory** consists of a very large number of electronic circuits, each circuit capable of storing a single binary digit (bit). It is convenient to organise the memory into groups, each group consisting of eight individual memory elements (1 byte). This is because most current microprocessors operate on an eight bit data word basis. Therefore each group in the memory is capable of storing a single microprocessor instruction, data word or calculation result. A section of the memory is reserved for storing the sequence of microprocessor instructions forming the program.

(c) Each location in a memory system must be assigned a unique identifying label. This enables the system to know *where* data is stored within memory and enables a particular data word to be recalled by specifying its memory location when retrieving it from memory. The identifying label is called an **address** and all memory systems have a number of conductors connected to them called **address lines**, which together form an **address bus**. Therefore, if a binary number representing a memory location is set up on the address bus, then that location is connected to the memory input/output lines.

(d) The **data bus** is responsible for carrying the data words between memory or I/O and the microprocessor and also in the reverse direction – hence it is a 'bi-directional data bus'. Each conductor in the data bus is responsible for carrying one bit of a data word. The data bus therefore has as many conductors in it as there are bits in the data word – 8 bits for an 8-bit microprocessor.

(e) The **control bus** is not, strictly speaking, a bus in the true sense, but a miscellaneous collection of microprocessor inputs and outputs which are responsible for overall control of the system and which, for convenience, are grouped together as a bus. Controls such as system reset and memory read/write are amongst those carried by this bus.

(f) The microcomputer is used as the processing part of a system. It was stated in Chapter 1 that systems have inputs and outputs, and the microcomputer controlled system is no exception. Devices connected to the microcomputer inputs/outputs are called peripherals and these may be divided into two groups:
   (i) those which provide input signals for the microcomputer, called input peripherals, and
   (ii) those which are operated by signals arriving from the outputs of the microcomputer, called output peripherals.

4 The contents of a memory may need to be of a fixed nature for some applications and variable for others. Therefore two types of memory are needed:
   (i) **ROM (read only memory)**. This is a type of memory whose contents are fixed during manufacture. The contents of this memory may be read by the user, but may not be altered, that is, the user cannot write information into a memory of this type.

(ii) **RAM (random access memory)**. This is more correctly called a read/write memory. Information may be read from memory and also written (stored) into memory. Therefore information contained in a RAM may be of a changeable nature. Whether a RAM is being read from or written to is determined by the state of the read/write line in the control bus.

When the power supply to a RAM is switched off, the contents are lost. This means that when the supply is restored, the contents of a RAM are completely random and unpredictable, and therefore the information must be reloaded. This is not so with a ROM, the contents of which are permanently stored so that information is available immediately after switching on the power supply. A microprocessor program may be stored in either ROM or RAM. Programs stored in RAM are called **software**, whilst programs stored in ROM are called **firmware**. If a RAM is used for program storage (software), then some means must be provided for loading the program into memory. This usually means that a short program must already be in the system at switch-on time, specifically for this purpose, that is, some firmware must always be present. The software may then be loaded into RAM from some form of peripheral equipment, for example a keyboard or disk drive.

If ROM is used for program storage (firmware), then the program is available and ready for use after switch-on. However, some RAM is still usually necessary to cater for any variable data. Therefore, most systems contain both ROM and RAM.

5 A memory map is a diagram which shows how the memory, I/O and peripheral devices are distributed within the available address space. Most microprocessors have a 16-bit address capability which enables $65,536_{10}$ different locations in the range $\emptyset\emptyset\emptyset\emptyset_{16} - FFFF_{16}$ to be specified. A typical memory map which shows how this address space is used is illustrated in *Figure 10*. The position of devices on a memory map may be changed by connecting their chip select inputs to different outputs on the address decoder; however, some form of permanent memory is required at locations corresponding to the addresses generated by a microprocessor immediately following reset.

6 Each microprocessor instruction consists of two parts, an **operator** and an **operand** (see *Figure 11*). An operator defines which **arithmetic, logic** or **data transfer** operation is carried out. An operand defines which data is used during an operation specified by an operator. The operation of a microprocessor is based upon a sequence of events known as a 'fetch-execute cycle'.

The fetch operation involves getting an instruction out of memory, and passing it along the data bus to the microprocessor. The execute operation involves carrying out the process defined by an instruction. The fetch-execute cycle may be summarised as follows:

(i)   **fetch** an instruction from memory and store it in the microprocessor's **instruction register**;

(ii) **decode** the instruction within the microprocessor to determine the nature of the operation specified by the instruction;

(iii) if necessary, fetch more data;

(iv) carry out (**execute**) the instruction.

This process may be studied by referring to *Figure 12(a) to (d)* which shows the fetch-execute cycle for a typical **load accumulator from memory** instruction, i.e.:

LD A, (1825H)   [3A 25 18]

Assuming that this instruction is located in memory addresses 1800H to 1802H, the sequence consists of the following four stages:

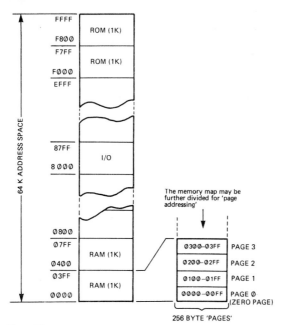

Figure 10

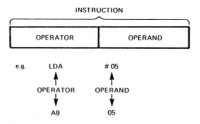

Figure 11

(i) Fetch the opcode (3A) from memory and transfer it into the instruction register (see *Figure 12(a)*),

(ii) Fetch the low byte (25) of the external address from memory, and transfer it to the lower eight bits of the MAR (see *Figure 12(b)*),

(iii) Fetch the high byte (18) of the external address from memory, and transfer it to the upper eight bits of the MAR (see *Figure 12(c)*) and,

(iv) Put the address in the MAR (1825) out onto the address bus, and transfer the data (05) from this address into the accumulator.

This sequence may be summarised as shown in *Table 2*.

Table 2

Step	Action
1	PC → address bus
2	PC = PC + 1
3	Data bus → IR
4	PC → address bus
5	PC = PC + 1
6	Data bus → MAR low
7	PC → address bus
8	PC = PC + 1
9	Data bus → MAR high
10	MAR → address bus
11	Data bus → accumulator

7 All operations in a microcomputer may be broadly classified as:
   (i) **Read** operations, during which a byte of data is transferred from one of the memory or I/O devices into the microprocessor;
   (ii) **Write** operations, during which a byte of data is transferred from the microprocessor to a memory or I/O device; and
   (iii) **Internal** operations, during which there is no activity on the system busses.

   The timing of data transfers between a microprocessor and its memory or I/O devices is critical, and may be studied by means of appropriate **timing diagrams**. A timing diagram shows, in correct time sequence, the states of all relevant bus and control lines during a specified operation. The data on any bus or control line may at any instant be either **valid**, **invalid**, or **uncertain**, examples of each of these conditions being shown in *Figure 13*.

   The timing of microprocessor read and write cycles depends very much upon the characteristics of the memory or I/O devices used.

8 The timing of a **memory read cycle** may be studied by reference to *Figure 14*. From this diagram it can be seen that a time delay exists between applying the address and control signals to the memory and the appearance of the selected data on the memory outputs. This is known as the memory **access time**, and is the combined effect of several different delays within the memory and varies considerably between different types of memory. A microprocessor must take into account the memory access time when reading a memory, and thus wait for a suitable time period before clocking data into its internal registers from the data bus. The actual clocking arrangements vary from one microprocessor to another, but a typical arrangement for a microprocessor which uses a two phase clock is shown in *Figure 15* (see *Worked problem 10*).

   It should be noted that the maximum clocking frequency for a microcomputer may be determined by the access time of the memory used, and that if a higher speed microprocessor is used, it may be necessary to use memory devices with a shorter access time to take advantage of this higher speed.

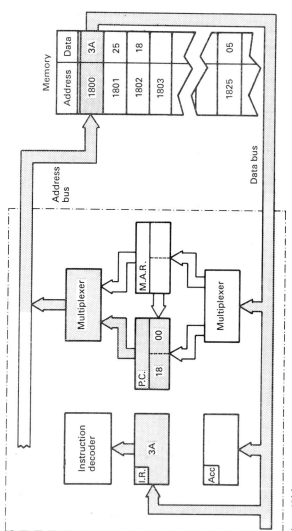

Figure 12(a)

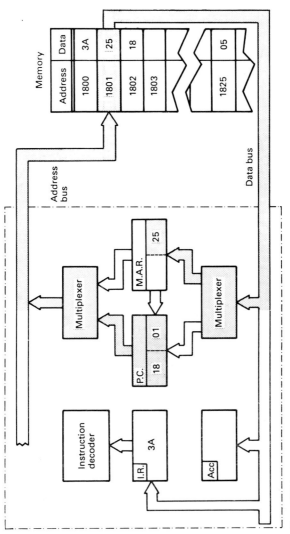

Figure 12(b)

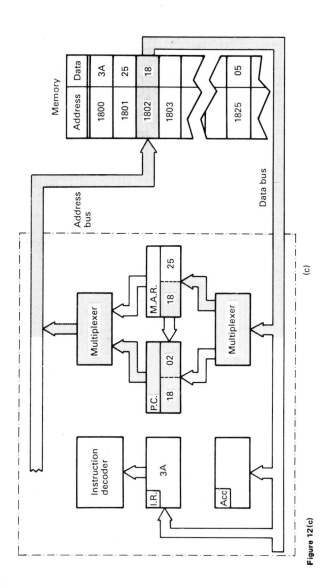

Figure 12(c)

56

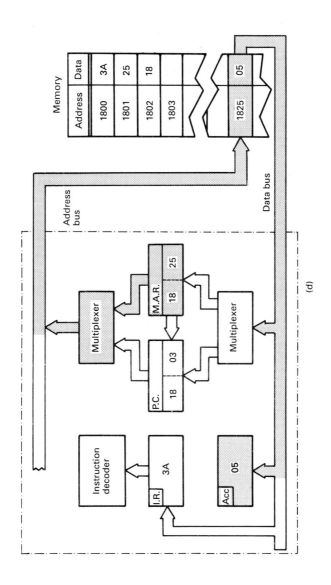

(d)

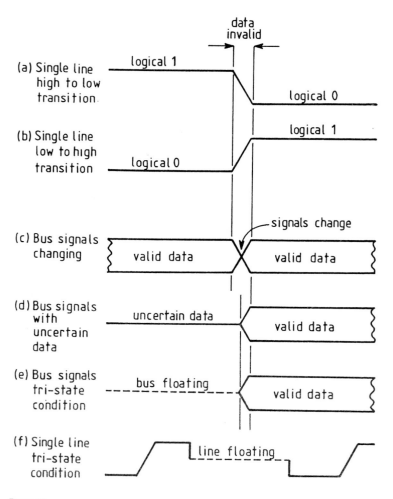

data
invalid

(a) Single line
high to low
transition.

logical 1

logical 0

(b) Single line
low to high
transition

logical 0

logical 1

signals change

(c) Bus signals
changing

valid data

valid data

(d) Bus signals
with
uncertain
data

uncertain data

valid data

(e) Bus signals
tri-state
condition

bus floating

valid data

(f) Single line
tri-state
condition

line floating

Figure 13

58

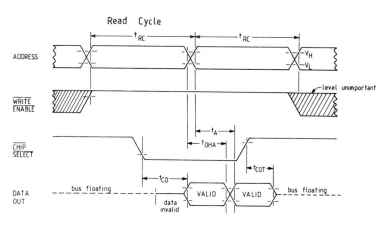

Read Cycle

t_{RC} = Read Cycle Time t_{OHA} = Output Hold from Address Change t_{COT} = Chip Select to Output Tri-State
t_A = Access Time t_{CO} = Chip Select to Output Valid V_H = Logical 1 level
    V_L = Logical 0 level

Figure 14

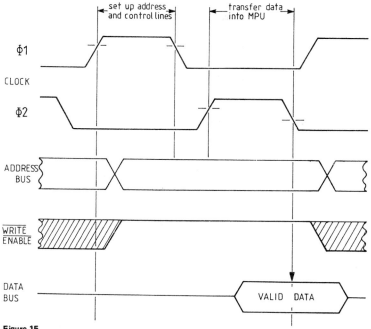

Figure 15

9 The timing of a **memory write cycle** may be studied by reference to *Figure 16*. In this case, the address, data and control lines must be held stable for a period of time long enough for the memory to respond and store the data in the selected location. The data bus may stabilise some time after setting up the address and control lines, but must then be held stable for a specified time to enable writing to take place. In a typical microprocessor which operates with a two phase clock, the address and control signals are set up during the first clock phase (Φ1), and data is stored in the memory during the second clock phase (Φ2). The memory used should be capable of completing the storage operation before the trailing edge of Φ2 as shown in *Figure 17*.

   This timing will influence the choice of memory and clocking frequency for a given application. The control lines necessary to support the memory read and write cycles, described above, are shown in *Figures 18(a) and (b)*.

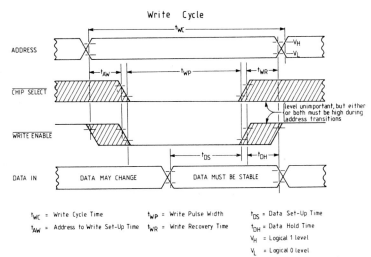

Write Cycle

t_{WC} = Write Cycle Time	t_{WP} = Write Pulse Width
t_{AW} = Address to Write Set-Up Time	t_{WR} = Write Recovery Time

t_{DS} = Data Set-Up Time
t_{DH} = Data Hold Time
V_H = Logical 1 level
V_L = Logical 0 level

Figure 16

60

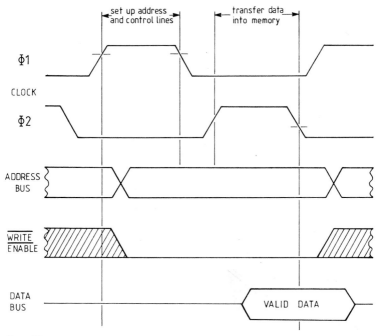

Figure 17

61

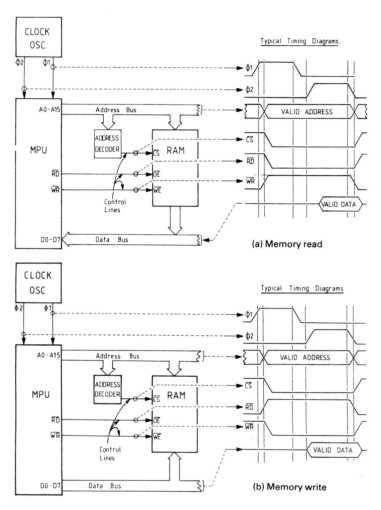

Figure 18

Problem 1   Describe the action of a typical bistable or flip-flop circuit which may be used to store a single bit of digital information.

A basic bistable or flip-flop circuit may be constructed from two cross-coupled NAND gates as shown in *Figure 19* (see *Appendix A* for details of NAND gates). This particular circuit is called an $\overline{S}$-$\overline{R}$ flip-flop. The action of this

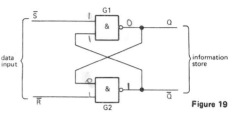

Figure 19

circuit may be described by considering the initial conditions to be:

$$\overline{S} = 1, \quad \overline{R} = 1, \quad Q = \emptyset, \quad \overline{Q} = 1$$

G1 has a logical 1 on both of its inputs and this maintains Q at a logical $\emptyset$. G2 has its inputs at logical $\emptyset$ and logical 1 respectively, and this maintains $\overline{Q}$ at logical 1.

These conditions are stable, therefore it can be considered that Q stores a logical $\emptyset$ (or $\overline{Q}$ stores a logical 1).

If S is changed to logical $\emptyset$, G1 inputs are now logical 1 and logical $\emptyset$ respectively, which causes Q to change to a logical 1. Since Q is connected to one input of G2, this gate now has a logical 1 on both its inputs. Therefore $\overline{Q}$ changes to logical $\emptyset$, which in turn is connected to one input of G1 to hold Q at logical 1.

This condition is steady, and is maintained even if S returns to logical 1. Therefore it can be considered that Q stores a logical 1 (or $\overline{Q}$ stores a logical $\emptyset$). Q may be changed back to a logical $\emptyset$ by the application of a logical $\emptyset$ to $\overline{R}$ which reverses the above process.

For data storage, it is not always convenient to have a circuit with two inputs. This problem may be overcome by combining $\overline{S}$ and $\overline{R}$ to form a **D-type flip-flop** (see *Figure 20*). Information to be stored is applied to the D input. Consider the conditions when Q is at logical $\emptyset$, and a logical 1 is applied to the D terminal:

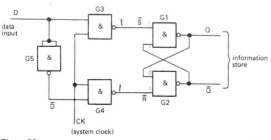

Figure 20

With **CK at logical 0;** G3 inputs are at logical 0 and logical 1 respectively, which causes G3 output to be maintained at logical 1. This holds $\overline{S}$ at logical 1. Due to the action of G5, G4 inputs are both at logical 0 which causes its output to be maintained at logical 1. This holds $\overline{R}$ at logical 1.

When CK changes to logical 1; G3 inputs are both at logical 1 which causes its output to change to logical 0. Hence, $\overline{S}$ becomes logical 0. G4 inputs are logical 1 and logical 0 respectively, which causes its output to remain at logical 1. Hence $\overline{R}$ remains at logical 1.

Since $\overline{S} = 0$ and $\overline{R} = 1$, Q changes to logical 1. If logical 0 is applied to D, the above process reverses and Q changes to logical 0. Thus, Q stores whatever data is applied to the D terminal at clocking time.

(Note that a system clock is a pulse generator which is responsible for timing and synchronizing all data transfers in a system.)

Problem 2 Describe typical circuits which may be used for binary addition.

The rules for binary addition may be summarised as shown by *Table 3*, assuming that no carry-in takes place (i.e. they are the rules for **two bit addition**). Observation of the '**SUM** output shows that it has the same logic pattern as when 'A' is **EX-ORed** with 'B'. Observation of the '**CARRY**' output shows that it has the same logic pattern as when 'A' is **ANDed** with 'B' (see *Appendix A*, page 246). Therefore, a binary adder may be constructed from a combination of one EX-OR gate and one AND gate, arranged as shown in *Figure 21*. This circuit is called a '**half-adder**' since it does not permit a carry-in from previous additions. For most microprocessor applications, a '**full-adder**', capable of adding three bits is required. This may be constructed from two half-adder circuits connected as shown in *Figure 22*. The action of this circuit may be studied by referring to *Table 4*.

Table 3

A	B	SUM	CARRY
0	0	0	0
0	1	1	0
1	0	1	0
1	1	0	1

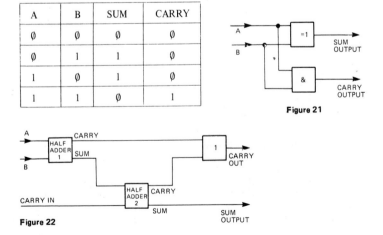

Figure 21

Figure 22

Table 4

A	B	CARRY-IN	SUM	CARRY-OUT
0	0	0	0	0
0	1	0	1	0
1	0	0	1	0
1	1	0	0	1
0	0	1	1	0
0	1	1	0	1
1	0	1	0	1
1	1	1	1	1

A **carry out** occurs if:
(i) A and B are both logical 1 (carry generated from half-adder number 1), or
(ii) carry-in is logical 1 and either A or B are logical 1 (carry generated from half-adder number 2).
This is taken care of by connecting the carry-out from both half-adders to an OR gate. The sum is obtained by adding the sum of A and B to the carry-in using half-adder number 2.

Problem 3 Explain the behaviour and uses of the following flags in the status register of a microprocessor:
(a) the N flag, (b) the Z flag, and (c) the C flag.

(a) An 'N' **(negative)** flag is a bistable, or flip-flop within a microprocessor, which is set ($N = 1$) automatically each time that a **negative result** is obtained as a consequence of a microprocessor executing certain instructions. Conversely, if an N-flag is reset, or cleared ($N = 0$), a positive result is indicated.
  Certain microprocessor instructions test the state of an N-flag to determine what action they should take (see 'branch' instructions, Chapter 4). Since microprocessors use two's complement arithmetic, an N-flag assumes the logic level of **bit 7** of the accumulator (or other flag-setting register).

(b) A 'Z' **(zero)** flag is a bistable, or flip-flop within a microprocessor, which is set ($Z = 1$) automatically each time that a result of **zero** is obtained as a consequence of a microprocessor executing certain instructions. Conversely, if a Z-flag is reset, or cleared ($Z = 0$), a non-zero result is indicated. Certain microprocessor instructions test the state of a Z-flag to determine what action to take (see 'branch' instructions, Chapter 4). Logically, a Z-flag indicates the result of NORing all of the bits of the accumulator (or other flag-setting register).

(c) A 'C' **(carry)** flag is a bistable, or flip-flop within a microprocessor which is set ($C = 1$) automatically each time that the result of an addition to the accumulator gives an answer of **greater magnitude** than can be contained in the accumulator. With an eight-bit accumulator, the C-flag is set whenever a result of greater magnitude than 255_{10} (FF_{16}) is obtained.
  Conversely, if the C-flag is reset, or cleared ($C = 0$), a result less than the maximum value which may be contained in the accumulator is indicated. During subtraction from the accumulator, a C-flag behaves as a 'borrow' flag. In this mode, its operation varies slightly between different microprocessors.

In some microprocessors, a C-flag is set (C = 1) if a borrow-in to bit 7 of the accumulator occurs (when a number larger than the contents of the accumulator is subtracted from the accumulator); in other microprocessors a C-flag is cleared (C = $\emptyset$) under the same conditions.

Problem 4   Explain the operation and uses of the V-flag in a microprocessor status register.

A V-flag is used to denote a **two's complement overflow**. It is set (V = 1) when the sign of a two's complement result is inadvertently changed due to its magnitude exceeding the maximum permitted, i.e. when a carry occurs from bit 6 to bit 7 in an 8-bit accumulator to give a result outside the range -128 to $+127$.

Therefore, a V-flag acts as a carry-out flag for signed arithmetic operations, since the C-flag becomes meaningless. If unsigned arithmetic is used, the V-flag may be ignored. If signed arithmetic is used, the V-flag may be used for **sign correction** purposes when necessary. Note that the operation of a microprocessor never changes between signed and unsigned arithmetic, only the interpretation placed on the MSB of the accumulator and the V-flag by the user changes. An overflow occurs:

(i)   when adding two **large positive** numbers,

(ii)   when adding two **large negative** numbers,

(iii)   when subtracting a **large positive** number from a **large negative** number, or

(iv)   when subtracting a **large negative** number from a **large positive** number.

An overflow **never** occurs when two numbers of **opposite sign** are added. The operation of a V-flag may be studied by considering the following calculations:

(i)

$$
\begin{array}{l}
\text{no} \quad \emptyset\ 1\ 1\ \emptyset\ \emptyset\ 1\ \emptyset\ \emptyset \quad (+1\emptyset\emptyset_{10}) \\
\text{carry} \quad \emptyset\ 1\ 1\ \emptyset\ \emptyset\ 1\ \emptyset\ \emptyset \quad (+1\emptyset\emptyset_{10})
\end{array}
\Big\} \text{ added}
$$

flags C $\boxed{\emptyset}$ $\leftarrow$ 1 1 $\emptyset$ $\emptyset$ 1 $\emptyset$ $\emptyset$ $\emptyset$ (-56_{10})

V $\boxed{1}$

carry b_6 to b_7 error   (should be $+2\emptyset\emptyset_{10}$)

(ii)

$$
\begin{array}{l}
1\ \emptyset\ \emptyset\ \emptyset\ 1\ \emptyset\ \emptyset\ \emptyset \quad (-12\emptyset_{10}) \\
\text{carry} \quad 1\ \emptyset\ \emptyset\ \emptyset\ 1\ \emptyset\ \emptyset\ \emptyset \quad (-12\emptyset_{10})
\end{array}
\Big\} \text{ added}
$$

flags C $\boxed{1}$ $\leftarrow$ $\emptyset$ $\emptyset$ $\emptyset$ 1 $\emptyset$ $\emptyset$ $\emptyset$ $\emptyset$ $(+16_{10})$

V $\boxed{1}$ no

carry b_6 to b_7 error   (should be $-24\emptyset_{10}$)

(iii)

$$
\begin{array}{l}
\text{no} \quad 1\ \emptyset\ \emptyset\ \emptyset\ 1\ \emptyset\ \emptyset\ \emptyset \quad (-12\emptyset_{10}) \\
\text{borrow} \quad \emptyset\ 1\ 1\ \emptyset\ \emptyset\ 1\ \emptyset\ \emptyset \quad (+1\emptyset\emptyset_{10})
\end{array}
\Big\} \text{ subtracted}
$$

flags C $\boxed{\emptyset}$ $\leftarrow$ $\emptyset$ $\emptyset$ 1 $\emptyset$ $\emptyset$ 1 $\emptyset$ $\emptyset$ $(+36_{10})$

V $\boxed{1}$ borrow b_6 to b_7 error   (should be $-22\emptyset_{10}$)

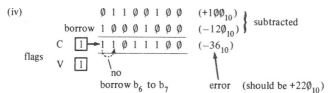

(iv)

$$
\begin{array}{l}
\phantom{\text{borrow }}\; \emptyset\;1\;1\;\emptyset\;\emptyset\;1\;\emptyset\;\emptyset \qquad (+1\emptyset\emptyset_{10}) \\
\text{borrow }\; 1\;\emptyset\;\emptyset\;\emptyset\;1\;\emptyset\;\emptyset\;\emptyset \qquad (-12\emptyset_{10})
\end{array} \Bigg\} \text{ subtracted}
$$

$$
\text{C }\boxed{1}\!\to\!1\;1\;\emptyset\;1\;1\;1\;\emptyset\;\emptyset \qquad (-36_{10})
$$

flags V $\boxed{1}$ no   borrow b_6 to b_7 error (should be $+22\emptyset_{10}$)

The above calculations are examples of all cases where an overflow may occur. From these examples it can be seen that an overflow occurs when:
(a) a carry/borrow occurs between bit 6 and bit 7 of the accumulator, but no carry/borrow between bit 7 of the accumulator and the C-flag, or
(b) no carry/borrow occurs between bit 6 and bit 7 of the accumulator, but a carry/borrow occurs between bit 7 of the accumulator and the C-flag.

Logically, the state of a V-flag may be expressed as the result of EX-ORing the carry between bit 6 and bit 7 of the accumulator with the carry from bit 7 of the accumulator to the C-flag.

Problem 5   Explain what is meant by the term 'multiple-precision arithmetic', and show how a C (carry) flag is used to enable this form of arithmetic to be performed.

With written arithmetic, the magnitude of numbers manipulated is limited only by the width of paper and size of writing used. In practice, these limitations impose virtually no restraint on the magnitude of numbers handled. With electronic devices, such as microprocessors, there are physical restraints on the magnitude of numbers which may be processed. For example, a typical microprocessor has an 8-bit accumulator, and therefore is unable to manipulate binary numbers which have more than eight bits (i.e. greater than 255_{10}). '**Multiple precision arithmetic**' refers to arithmetic involving numbers which are greater in magnitude than that permitted by an accumulator. Consider the addition of two 24-bit numbers when using an 8-bit microprocessor. Clearly this must involve three separate additions of three pairs of 8-bit numbers. The C-flag acts as a **link** between the three additions, acting as a **carry-out store** for one addition, and providing a **carry-in** for the next addition. This process may be illustrated by using the addition of two 24-bit binary numbers, $\emptyset 111\emptyset 1\emptyset \emptyset 11\emptyset \emptyset 111\emptyset 1\emptyset \emptyset \emptyset 11\emptyset 1_2$ ($765 5\emptyset 53_{10}$) and $1\emptyset \emptyset 1\emptyset 11\emptyset 111\emptyset \emptyset 111111\emptyset \emptyset 1\emptyset 1\emptyset 1_2$ (9889685_{10}), as an example:

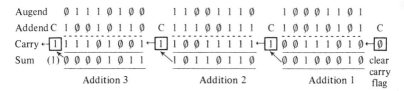

Each of the three additions are similar to those already encountered in Chapter 2. Initially the carry flag must be cleared to enable the first addition to take place, but thereafter its state is determined by the result of each addition, thus acting as a carry link between successive additions. A C-flag may be used in a similar manner to enable multiple precision subtraction to be implemented.

Problem 6 State what is meant by the term 'binary coded decimal' (BCD) and explain how it is possible for a microprocessor to perform BCD arithmetic.

For the convenience of human operators, most microprocessor controlled machines are designed to accept decimal inputs from, for example, a keyboard and deliver decimal outputs to, for example, seven-segment displays. In order to facilitate this form of operation, a special form of binary notation is used within a micro-processor called '**binary coded decimal**' (**BCD**). Using this notation, each decimal digit is coded, in isolation, into its binary equivalent. For example, 9367_{10} is coded as follows:

9	3	6	7
1001	0011	0110	0111

Thus, 9367_{10} has a BCD equivalent of 1001001101100111_2.

There are sixteen possible combinations available when using four bits in a binary number, but a decimal system has only ten discrete digits. Therefore this system has many unused binary combinations (redundancy).

A microprocessor generally deals with BCD numbers in one of three ways:
(i)   by using **special instructions** for decimal add and subtract;
(ii) by using a '**decimal adjust**' instruction after each arithmetic operation, or
(iii) by using a '**decimal mode**' instruction which enables a microprocessor to operate in a BCD mode until cancelled by a 'clear decimal mode' instruction.

In all three cases, conversion into BCD form takes place by the microprocessor inspecting each hexadecimal digit (four bits) in its accumulator, and **adding 6** to a digit if its value **exceeds 9**. For example, consider the BCD addition of 38_{10} to 25_{10}.

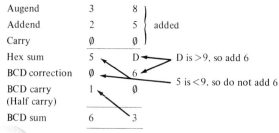

Note, adding 6 to D gives a result of 3, carry 1. This carry is sometimes called a '**half-carry**', and a flag (**H-flag**) is included in some status registers to indicate the fact that this type of carry has occurred. An H-flag is not accessible to a micro-processor user, however.

A microprocessor must be capable of translating instructions obtained from memory into suitable control signals necessary for carrying out those instructions. A microprocessor must also generate addresses for external memory and internal registers used by each instruction, provide the necessary inputs for the ALU and control the microprocessor's internal busses so that each instruction may be properly executed. For example, consider the 65Ø2 instruction '**transfer accumulator to X' (TAX)**, in which the contents of the accumulator are transferred to an 8-bit index register, X. The steps necessary to carry out this instruction are illustrated in *Figure 23(a) to (e)*.

Ordinary logic circuits may be used to perform the above tasks. A microprocessor which uses this form of construction is said to be '**hard wired**'. An alternative approach is to have a **control system** within the main control section of a microprocessor. This secondary control section is capable of executing instruction steps which are stored in **read-only memory (ROM)** within a microprocessor.

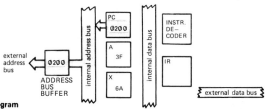

Figure 23 (a) Copy program counter (PC) into address bus buffer and put address onto external address bus

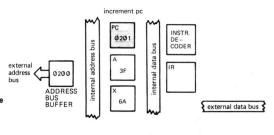

Figure 23 (b) Break the path between the program counter and the address bus buffer and increment the program counter

69

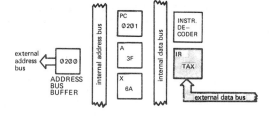

Figure 23 (c) Fetch an instruction (TAX) from the address specified by the address bus buffer contents and store in the instruction register (IR)

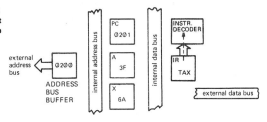

Figure 15 (d) Decode the instruction and as a result open a data path between the accumulator (A) and index register (X)

Figure 15 (e) Copy the data in the accumulator (A) into the index register (X)

These instruction steps are very elementary in nature and are called '**micro-instructions**'. A sequence of microinstructions necessary to form each microprocessor instruction is called a '**microprogram**'. The microinstructions necessary to perform the TAX instruction shown in *Figure 23* are as follows:

(i) **Microinstruction 1**: put the contents of the program counter (PC) onto the address bus;

(ii) **Microinstruction 2**: increment the program counter;

(iii) **Microinstruction 3**: transfer the signals on the external data bus into the instruction register (IR);

(iv) **Microinstruction 4**: decode the instruction.

(v) **Microinstruction 5**: copy the contents of A into X.

(Microinstructions 1 and 2 are common to all instructions.)

The use of microprogramming allows greater flexibility in microprocessor design, but results in a slower operating speed. Some more advanced microprocessors are microprogrammable, and this permits a system designer to specify his own instruction set.

Note: a microprogram should not be confused with a microprocessor program.

A microprocessor program is written using instructions which are predetermined by microprograms stored within a microprocessor, and are usually built in during manufacture.

Problem 8 With the aid of diagrams, describe the fetch-execute cycle for a 'jump unconditional' instruction.

The fetch-execute cycle for a typical 'jump unconditional' instruction is illustrated in *Figure 24(a) to (d)*. The following instruction is used as an example:

  JP 1825H   [C3 25 18]

and, assuming that this instruction is located in memory addresses 1800H to 1802H, the fetch-execute cycle is:

(i)   Fetch the opcode (C3) from memory and transfer it into the instruction register (see *Figure 24(a)*).
(ii) Fetch the low byte (25) of the jump destination address from memory, and transfer it to the lower eight bits of the MAR (see *Figure 24(b)*).
(iii) Fetch the high byte (18) of the jump destination address from memory, and transfer it to the upper eight bits of the MAR (see *Figure 24(c)*).
(iv) Transfer the address in the MAR (1825) into the PC (see *Figure 24(d)*).

This sequence may be summarised as shown in *Table 5*.

Table 5

Step	Action
1	PC → address bus
2	PC = PC + 1
3	Data bus → IR
4	PC → address bus
5	PC = PC + 1
6	Data bus → MAR low
7	PC → address bus
8	PC = PC + 1
9	Data bus → MAR high
10	MAR → PC

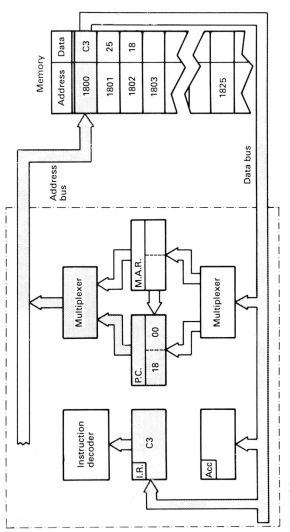

Figure 24(a)

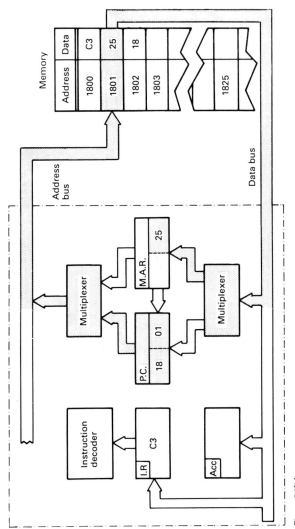

Figure 24(b)

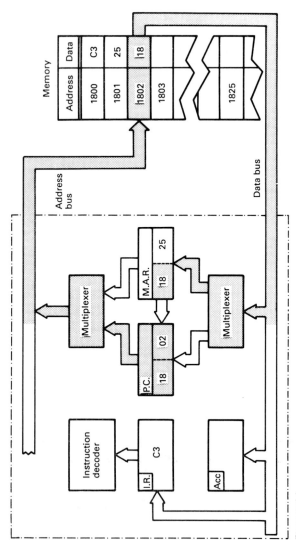

Figure 24(c)

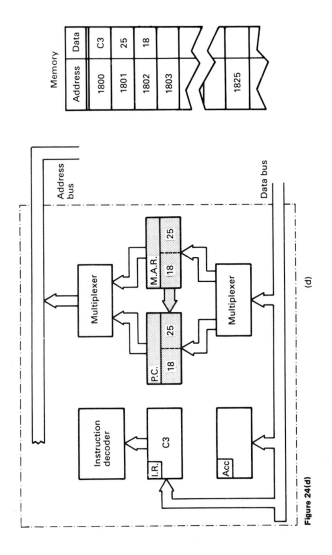

Figure 24(d)

With reference to *Figure 25*, state the function of the following control pins:
(i) R/W; (ii) $\overline{\text{RES}}$; (iii) SYNC; (iv) RDY.

(i)   The **R/W (read/write) line** allows the microprocessor to control the **direction** of **data transfers** between it and its support chips. This line is high (logical 1) except when the microprocessor is writing to memory or to a PIA. All transitions on this line occur during phase one of the clock.

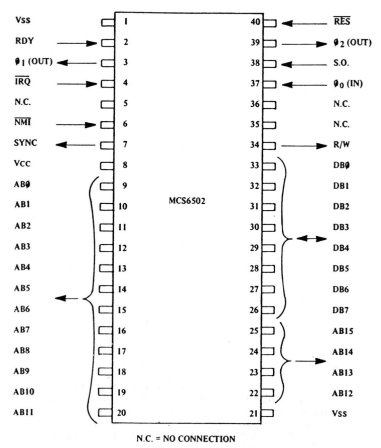

N.C. = NO CONNECTION

Figure 25 MCS6502 pinout designation

(ii) **RES** is a **reset** input and is used to initialize the microprocessor from a power down condition or to restart a program if the system is already in operation.

(iii) **SYNC** is an output signal which is used to identify those cycles during which the microprocessor is carrying out an **opcode fetch**. The SYNC line goes high (logical 1) during clock phase one of an opcode fetch, and stays high for the remainder of that cycle. This signal may be used in conjunction with the $\overline{NMI}$ input to enable single instruction execution (single step) to be implemented.

(iv) The **RDY (ready)** input delays the execution of any cycle during which RDY is pulled low (logical $\emptyset$). This input should change state during a phase one clock cycle. The change is recognised during the next phase two clock cycle and enables or disables execution of the current internal machine cycle. The main function of the RDY input is to delay execution of a program fetch cycle until data is available from memory, thus enabling the 6502 to operate with slow access memories without requiring a reduction in the clock frequency.

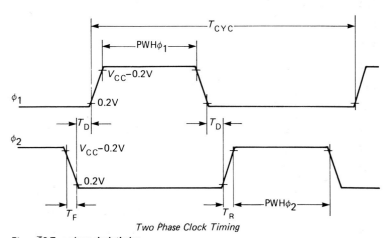

Two Phase Clock Timing

Figure $\overline{2}6$ Two-phase clock timing

A microprocessor requires a **continuous clock waveform** which is used to control all signal transitions within the system. The 6502 microprocessor requires a **non-overlapping, two phase clock** generator circuit which delivers signals of the type shown in *Figure 26*. The two phases are known as **Φ1** and **Φ2**, and the microprocessor is arranged such that **address changes** take place during **Φ1**, and **data transfers** take place during **Φ2**.

The timing for reading data from memory or peripheral devices is shown in *Figure 27*.

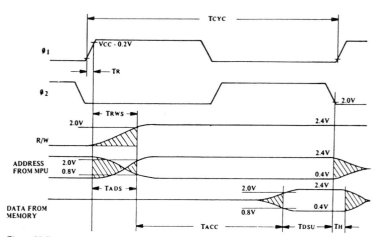

Figure 27 Timing for reading data from memory or peripherals (6502)

Changes in state of the address and R/W lines of a 6502 microprocessor are initiated by the rising edge of its Φ1 clock pulse. The R/W line stabilises after a short time period T_{RWS} (**read/write set up time**), and a stable address is obtained after a time period T_{ADS} (**address set up time**). In response to these signals, the memory or peripheral device puts its information out onto the data bus, and this becomes stable after a time period T_{ACC} (**memory or peripheral read access time**). This information must be held stable on the data bus for a time period T_{DSU} (**data stability time period**) prior to the falling edge of the Φ2 clock pulse, and for a time period T_H (**data hold time**) after the falling edge of Φ2 to enable the data to be read into the microprocessor on the falling edge of Φ2.

Typical times for this sequence when using a 1 MHz clock are shown in *Table 6*.

The timing for writing data to memory or peripheral devices is shown in *Figure 28*.

Changes in state of the address and R/W lines of a 6502 microprocessor are initiated by the rising edge of its Φ1 clock pulse. The R/W line stabilises after a short time period T_{RWS} (**read/write set up time**), and a stable address is obtained after a time period T_{ADS} (**address set up time**). The transfer of information from the microprocessor is initiated by the rising edge of its Φ2 clock pulse, and stable data appears on the data bus after a short time period T_{MDS} (**data set up time from MPU**). This information must be held stable on the data bus until a short time after the falling edge of Φ2, T_H (**data hold time**), to allow time for the memory or peripheral device to read the data.

Typical times for this sequence when using a 1 MHz clock are shown in *Table 6*.

Table 6 Clock and read/write timing table (1 MHz operation)

CHARACTERISTIC	SYMBOL	MIN.	TYP.	MAX.	UNIT
Cycle Time	T_{CYC}	1.0 us	--	--	usec
Clock Pulse Width Ø1 (Measured at Vcc-0.2v) Ø2	PWH Ø1 PWH Ø2	430 430	--	--	nsec
Rise and Fall Times (Measured from 0.2V to Vcc-0.2V)	T_F, T_R	--	--	25	nsec
Delay time between Clocks (Measured at 0.2V)	T_D	0	--	--	nsec

CHARACTERISTIC	SYMBOL	MIN.	TYP.	MAX.	UNIT
Read/Write Setup Time from MCS650X	T_{RWS}	--	100	300	ns
Address Setup Time from MCS650X	T_{ADS}	--	200	300	ns
Memory Read Access Time T_R $T_{CYC} - (T_{ADS} - T_{DSU} - tr)$	T_{ACC}	--	--	500	ns
Data Stability Time Period	T_{DSU}	100	--	--	ns
Data Hold Time	T_H	10	30	--	ns
Enable High Time for DBE Input	T_{EH}	430	--	--	ns
Data Setup Time from MCS650X	T_{MDS}		150	200	ns

Problem 13   The pin-out designation of a Z80 microprocessor is shown in *Figure 29*.

With reference to *Figure 29*, state the function of each of the following control pins: (i) M̄1̄; (ii) M̄R̄ĒQ̄; (iii) Ī̄Ō̄R̄Q̄; (iv) R̄D̄; (v) W̄R̄.

(i)   **M̄1̄** is an **active low output** and indicates that the current machine cycle is an opcode fetch cycle (**machine cycle 1**). Note that during the execution of 2-byte opcodes, M̄1̄ is generated as each opcode byte is fetched. M̄1̄ also occurs with Ī̄Ō̄R̄Q̄ to indicate an **interrupt acknowledge cycle**.

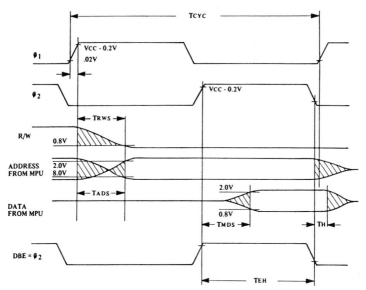

Figure 28 Timing for writing data to memory or peripherals

(ii) $\overline{\text{MREQ}}$ is an **active low memory request** signal and indicates that the address bus holds a **valid address** for a **memory read** or **memory write** operation.

(iii) $\overline{\text{IORQ}}$ is an **active low input/output request** signal, and indicates that the lower half of the address bus holds a **valid I/O address** for an **I/O read** or **write** operation.

(iv) $\overline{\text{RD}}$ is an **active low memory read signal**, and indicates that the microprocessor wants to read data from memory or an I/O device. The addressed memory or I/O device should use this signal to gate data onto the data bus.

(v) $\overline{\text{WR}}$ is an **active low memory write signal**, and indicates that the microprocessor data bus holds **valid data** to be stored in the addressed memory or I/O device.

Problem 14 With the aid of a suitable waveform diagram, describe the nature of the clock signal required by a Z80 microprocessor.

A Z80 microprocessor executes all instructions by stepping through a very precise set of a few basic operations which include the following:
(a) **memory read or write**;
(b) **I/O device read or write**; or
(c) **interrupt acknowledge**.

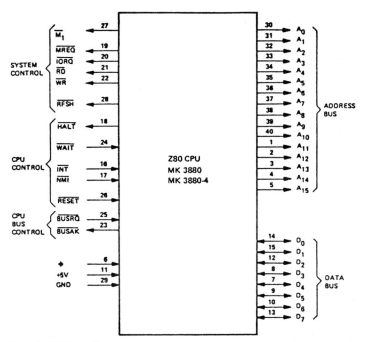

Figure 29 Z80 pin configuration

All instructions consist of a series of these basic operations. A **single phase clock** is used, and each basic operation can take from three to six clock periods to complete. Each clock period is referred to as a **'T' state**, and each basic operation is referred to as an **'M' (machine) cycle**. The use of clock cycles in relation to M cycles is shown in *Figure 30*.

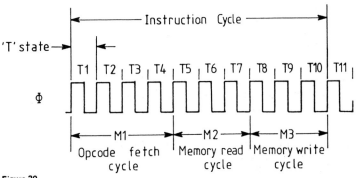

Figure 30

81

The timing diagram for a Z80 opcode fetch ($\overline{M1}$) cycle is shown in *Figure 31*.

The **program counter** contents are put out onto the **address bus** following the **rising edge** of the first clock cycle, **T1**. One half clock period later the $\overline{MREQ}$ and $\overline{RD}$ lines go **active** to indicate that the address bus is stable, and that the memory should put the opcode out onto the data bus. The microprocessor clocks the opcode into its instruction register from the data bus on the **rising edge of T3**. The rising edge of T3 is also used by the microprocessor to deactivate the **MREQ** and $\overline{RD}$ signals. During clock periods **T3** and **T4**, the microprocessor is **decoding** and **executing** the fetched instruction. Also, during clock periods T3 and T4, the lower seven bits of the address bus contain a memory refresh address, and the $\overline{RFSH}$ output becomes active to indicate that a refresh read of all dynamic memories should take place. The $\overline{RD}$ signal does not become active during this period to prevent data from all memory devices being simultaneously gated onto the data bus. The $\overline{MREQ}$ signal does, however, become active during T3 and T4, since it is only during the $\overline{MREQ}$ active time that the refresh address is guaranteed to be stable.

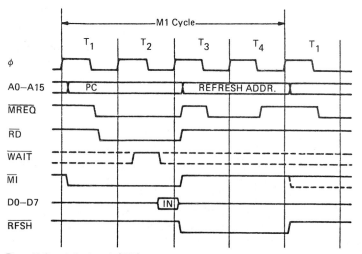

Figure 31 Opcode fetch cycle (Z80)

82

The timing diagram for a Z80 memory data read is shown in *Figure 32*.

The memory read cycle takes up three clock cycles rather than the four clock cycles required for an opcode fetch cycle, and there is no memory refresh operation involved. The generation of the address bus, $\overline{\text{MREQ}}$ and $\overline{\text{RD}}$ signals follow the same pattern as that used during the opcode fetch cycle. The data is clocked into the microprocessor on the **falling edge** of T3, however, and this edge of T3 is also used to deactivate $\overline{\text{MREQ}}$ and $\overline{\text{RD}}$ during a memory read cycle.

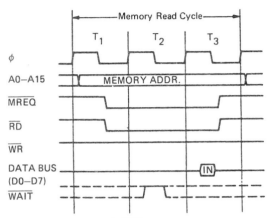

Figure 32 Memory read cycle (Z80)

Problem 17 With the aid of a timing diagram, describe the sequence of events which occur when a Z80 microprocessor performs a memory write cycle.

The timing diagram for a Z80 memory write cycle is shown in *Figure 33*.

The generation of an address for a memory write operation is initiated by the **rising edge** of **T1**. The address bus is allowed a half clock period in which to stabilise, then $\overline{\text{MREQ}}$ becomes active on the **falling edge** of **T1**. The transfer of data from the microprocessor to the data bus (and hence to the memory device) also occurs on the **falling edge** of **T1**. This data is allowed to stabilise for a half clock period, and the $\overline{\text{WR}}$ line becomes active on the **falling edge** of **T2**. The $\overline{\text{WR}}$ line may be used directly as a R/W pulse for virtually all types of memory device. In addition, the $\overline{\text{WR}}$ and $\overline{\text{MREQ}}$ lines becomes inactive on the **falling edge** of **T3**, which is a half clock period before the address and data bus signals change, thus satisfying the overlap requirements of practically all types of semiconductor memory.

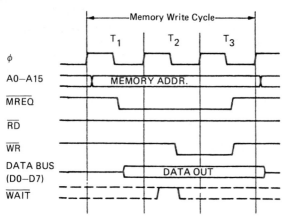

Figure 33 Memory write cycle (Z80)

Problem 18   The pin-out designation of a 6800 microprocessor is shown in *Figure 34*.
   With reference to *Figure 34*, state the function of the following control pins: (i) R/W; (ii) VMA; (iii) Halt; (iv) TSC; (v) DBE; (vi) BA.

(i)   The R/W (read/write) line allows the microprocessor to control the direction of data transfers between it and its support chips. This line is high (logical 1) except when the microprocessor is writing to memory or to a PIA or when placed in a high impedance state by one of the other control signals.

(ii)   The address bus of a 6800 microprocessor does not always hold a valid address for data transfers. A VMA signal output is high (logical 1) whenever the address bus does hold a valid address, and may be used to enable memory or peripheral interface devices.

(iii)   Halt is used to halt all activity in the microprocessor. If the Halt input is taken to a logical Ø level, the microprocessor stops at the end of an instruction, takes the BA output to a logical 1 level and VMA to a logical Ø level, and puts all tri-state lines into their high impedance condition.

(iv)   The TSC (three state control) input causes all of the address lines and the R/W line to go to a high impedance state when activated. This state occurs 700 ns after TSC = 2.0 V and is used mainly for DMA (direct memory access) applications.

(v)   The DBE (data bus enable) input is
      normally held in the high (logical 1)
      condition to enable the data bus drivers.
      For applications in which other devices
      take control of the data bus e.g. DMA
      (direct memory access), this input is held
      in the low (logical ∅) state.

(vi) The BA (bus available) output is
      normally held in the low (logical ∅) state.
      When activated, this output goes to a
      high (logical 1) state to indicate that the
      microprocessor has stopped, and that the
      address bus is available. This condition
      occurs if the H̅a̅l̅t̅ line is activated, or if
      the microprocessor is in the WAIT
      condition as a result of executing a 'WAI'
      instruction.

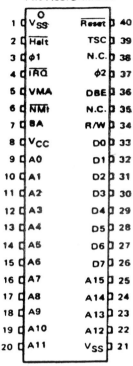

Figure 34 6800 pin assignment

Problem 19   With the aid of suitable waveform diagrams, describe the
nature of the clock signals required by a 6800 microprocessor.

A microprocessor requires a **continuous clock waveform** which is used to control
all signal transitions within the system. The 6800 microprocessor requires a **non-
overlapping, two phase clock** generator circuit which delivers signals of the type
shown in *Figure 35*. The two phases are known as Φ1 and Φ2, and the microprocessor is arranged such that **address changes** take place during Φ1, and **data
transfers** take place during Φ2.

Problem 20   With the aid of a timing diagram, describe the sequence of
events which occur when a 6800 microprocessor reads data from a memory
or peripheral device.

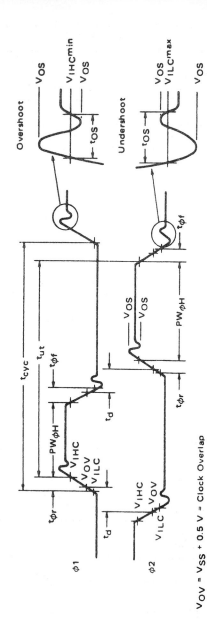

$V_{OV} = V_{SS} + 0.5\ V$ = Clock Overlap

measurement point

Figure 35

The timing for reading data from memory or peripheral devices is shown in *Figure 36*. Changes in state of the address, R/W and VMA lines of a 6800 microprocessor are initiated by the rising edge of its Φ1 clock pluse. These lines all become stable after a short time period t_{AD} (address delay time), and remain so until the falling edge of the following Φ2 clock pulse.

In response to these signals, the memory or peripheral device puts its information out onto the data bus, and this becomes stable after a time period t_{acc} (memory or peripheral read access time). This information must be held stable on the data bus for a time period t_{DSR} (data set up time, read) prior to the falling edge of the Φ2 clock pulse, and for a time period t_H (output data hold time) after the falling edge of Φ2 to enable the data to be read into the microprocessor on the falling edge of Φ2.

Typical times for this sequence when using a 1 MHz clock are shown in *Table 7*.

Problem 21 With the aid of a timing diagram, describe the sequence of events which occur when a 6800 microprocessor writes data to a memory or peripheral device.

The timing for writing data to memory or peripheral devices is shown in *Figure 37*. Changes in the state of the address, R/W and VMA lines of a 6800 microprocessor are initiated by the rising edge of its Φ1 clock pulse. These lines all become stable after a short time period t_{AD} (address delay time), and remain so until the falling edge of the following Φ2 clock pulse. The transfer of information from the microprocessor is initiated by the rising edge of the DBE (data bus enable) signal. In most applications the DBE input is connected to the Φ2 clock signal, therefore stable data appears on the data bus a short time period t_{DDW} (data delay time,

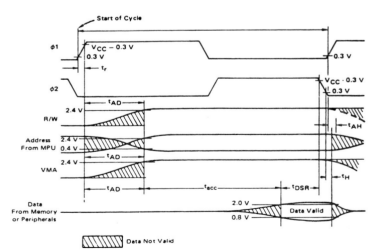

Figure 36 Memory read cycle (6800)

write) after the rising edge of the Φ2 clock pulse. This information must be held stable on the data bus until a short time after the falling edge of Φ2, t_H (output data hold time), to allow time for the memory or peripheral device to read the data; this timing changes when DBE is not connected to Φ2, as shown in *Figure 37*. Typical times for this sequence when using a 1 MHz clock are shown in *Table 7*.

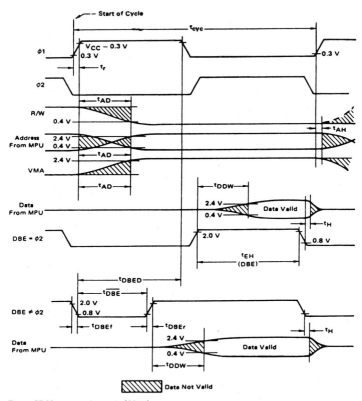

Figure 37 Memory write cycle (6800)

Table 7

Characteristic	Symbol	Minimum	Typ	Maximum	Unit
Address delay	t_{AD}	–	220	300	ns
Peripheral read access time $t_{acc} = t_{ut} - (t_{AD} + t_{DSR})$	t_{acc}	–	–	540	ns
Data setup time (Read)	t_{DSR}	100	–	–	ns
Input data hold time	t_H	10	–	–	ns
Output data hold time	t_H	10	25	–	ns
Address hold time (address, R/W, VMA)	t_{AH}	50	75	–	ns
Enable high time for DBE input	t_{EH}	450	–	–	ns
Data delay time (write)	t_{DDW}	–	165	225	ns
Processor controls					
Processor control setup time	t_{PCS}	200	–	–	ns
Processor control rise and fall time	t_{PCr}, t_{PCf}	–	–	100	ns
Bus available delay	t_{BA}	–	–	300	ns
Three state enable	t_{TSE}	–	–	40	ns
Three state delay	t_{TSD}	–	–	700	ns
Data bus enable down time during $\phi 1$ up time (Figure 37)	t_{DBE}	150	–	–	ns
Data bus enable delay (Figure 37)	t_{DBED}	300	–	–	ns
Data bus enable rise and fall times (Figure 37)	t_{DBEr}, t_{DBEf}	–	–	25	ns

C FURTHER PROBLEMS ON MICROPROCESSOR-BASED SYSTEMS

(a) SHORT ANSWER PROBLEMS

1 A group of bistable or flip-flop circuits, used for the temporary storage of data within a microprocessor are known collectively as . . . .

2 An individual bistable or flip-flop circuit, used by a microprocessor to indicate a particular internal condition, is known as a . . . .

3 A microprocessor has a group of individual bistable or flip-flop circuits to indicate various internal conditions which are collectively known as a . . . .

4 A binary counter, whose function within a microprocessor is to determine the sequence in which instructions are executed is known as a . . . .

5 The results of arithmetic operations are stored within a microprocessor in the . . . .

6 The section of a microprocessor where binary addition takes place is called the . . . .

7 The sequence of minute steps which enable a microprocessor to carry out a single instruction is called a . . . .

8 The part of an instruction which tells a microprocessor what process to carry out is called an . . . .

9 The part of an instruction which tells a microprocessor what piece of data to manipulate is called an . . . .

10 If the N-flag in a microprocessor is reset by an arithmetic operation, a . . . . result is indicated.

11 If the addition of two positive numbers in a microprocessor causes a negative answer to be produced, the . . . . will be set.

12 A circuit, which is constructed from an AND gate and an EX-OR gate with their inputs connected in parallel, may be used as a . . . .

13 An H-flag is used internally by a microprocessor when it is performing . . . . arithmetic.

14 Arithmetic which involves numbers of greater magnitude than can be contained in a microprocessor accumulator is known as . . . . arithmetic.

15 A part of a microprocessor which may be used to modify the address to which an instruction would otherwise refer is known as an . . . .

16 All machine cycles in a microprocessor may be broadly classified as . . . . , ... or . . . . operations.

17 Machine cycles during which a byte of data is transferred from a memory or I/O device into the microprocessor are known as . . . .

18 Machine cycles during which a byte of data is transferred from a microprocessor to a memory or I/O device are known as ....

19 Machine cycles during which there is no activity on the system busses are known as ....

20 The states of all relevant bus and control lines of a microprocessor may be studied in correct time relationship by means of a ....

21 The data on a bus or control line may, at any instant, be classified as .... , ... or ....

22 In *Figure 38*, point (a) shows that the bus ....

23 In *Figure 38*, point (b) shows that the bus ....

24 In *Figure 38*, point (c) shows that the bus ....

25 In *Figure 38*, point (d) shows that the bus ....

26 For a given frequency clock oscillator, the timing of microprocessor read and write cycles depends upon the characteristics of the ....

27 The time delay between applying address and control signals to a memory device and actually obtaining the data from the addressed location is known as the memory ....

28 A microprocessor clock oscillator is responsible for controlling all .... within a microcomputer.

29 In a microprocessor which uses a two phase clock, ... are set up during phase one of the clock, and .... occur during phase two of the clock.

30 Each clock cycle in a Z80 microprocessor is known as a ....

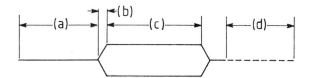

Bus signal

Figure 38

(b) MULTI-CHOICE PROBLEMS (answers on page 266)

In Problems 1 to 10 select the correct answer from those given.

1 If the result of an arithmetic operation is $\geq \emptyset$, the state of the N and V-flags in a microprocessor status register may be expressed as:
(a) $N \oplus V = \emptyset$;
(b) $N + V = \emptyset$;
(c) $N + V = 1$;
(d) $N \oplus V = 1$.

2 If the result of an arithmetic operation is $> \emptyset$, the state of the Z,N and V-flags in a microprocessor status register may be expressed as:
(a) $Z + (N \cdot V) = 1$;
(b) $Z \cdot (N \cdot V) = 1$;
(c) $Z + (N \oplus V) = \emptyset$;
(d) $Z \cdot (N \oplus V) = \emptyset$.

3 If the result of an arithmetic operation is $< \emptyset$, the state of the N and V flags in a microprocessor status register may be expressed as:

(a) $N \oplus V = 1$;
(c) $N \cdot V = 1$;

(b) $N \cdot V = \emptyset$;
(d) $N \oplus V = \emptyset$.

4 When two numbers are added in an 8-bit microprocessor, a carry-in to bit 7 and a carry out from bit 7 occurs. The state of the V and C-flags are as follows:

(a) $V = \emptyset$, $C = \emptyset$;
(c) $V = 1$, $C = 1$;

(b) $V = \emptyset$, $C = 1$;
(d) $V = 1$, $C = \emptyset$.

5 The V-flag in a microprocessor is never set if:

(a) two positive numbers are added;
(b) a large negative number is subtracted from a large positive number;
(c) a positive number is added to a negative number;
(d) a large positive number is subtracted from a large negative number.

6 A register, within a microprocessor, which is incremented by 1 before each operator or operand is fetched from memory, is called:

(a) an instruction register;
(c) an index register;

(b) a status register;
(d) a program counter.

7 After each instruction fetch, the next operation performed by a microprocessor is:

(a) to decode the instruction;
(c) to execute the instruction;

(b) to fetch more data;
(d) to fetch the next instruction.

8 When a microprocessor executes an 'add' instruction, the result of this addition is stored in:

(a) an external memory location;
(c) the accumulator;

(b) the ALU;
(d) an index register.

9 A half-adder circuit may be constructed by connecting together:

(a) two NAND gates with their inputs connected in parallel;
(b) an EX-OR gate and an AND gate with their outputs connected in parallel;
(c) two NAND gates with their outputs connected in parallel;
(d) an EX-OR gate and an AND gate with their inputs connected in parallel.

10 If the result of an arithmetic operation is $\leqslant \emptyset$, the state of the Z,N and V-flags in a microprocessor status register may be expressed as:

(a) $Z + (N \oplus V) = 1$;
(c) $Z + (N \cdot V) = 1$;

(b) $Z + (N \cdot V) = \emptyset$;
(d) $Z + (N \oplus V) = \emptyset$;

(c) CONVENTIONAL PROBLEMS

1 Explain the difference between a microprocessor which is **memory oriented** and a microprocessor which is **register oriented**.

2 Describe the **fetch-execute cycle** for a 'store accumulator in memory (STA)' instruction.

3 Show how **multiple-precision subtraction** may be performed using an 8-bit microprocessor, and state clearly the function of the C-flag in this process.

4 Draw a block diagram of a simple microprocessor, and indicate the function of each block.

5 State how the following operations may be performed on an 8-bit microprocessor, and include any flags tested in your answer:
   (a) checking whether a number in the accumulator is odd or even;
   (b) checking whether bit 4 is logical 1 or logical $\emptyset$;
   (c) complementing the number in the accumulator on a microprocessor which has no 'complement' instruction;
   (d) checking whether bit 7 is logical 1 or logical $\emptyset$.

6 Explain the advantages and disadvantages of using **binary coded decimal (BCD)** representation in a microprocessor. State how BCD arithmetic may be implemented on typical microprocessors.

7 The magnitudes of two two's complement numbers are compared in a microprocessor by subtracting one from another. Explain, using examples, why it is necessary to use the **C-flag** rather than the N-flag to indicate which is the larger of the two.

8 With the aid of a timing diagram, explain why the maximum clock frequency used with a microprocessor may be determined by the choice of memory devices.

9 With the aid of a timing diagram, describe the nature of a 'two phase clock' and explain how such a clock is used to control operations in a microprocessor.

10 With the aid of a timing diagram, show the sequence of machine cycles which occur when a 6502 microprocessor encounters a 'load accumulator immediate' instruction in its program.

11 With the aid of a timing diagram, show the sequence of machine cycles which occur when a Z80 microprocessor encounters a 'load accumulator immediate' instruction in its program.

12 With the aid of a timing diagram, show the sequence of machine cycles which occur when a 6800 microprocessor encounters a 'load accumulator immediate' instruction in its program.

13 With the aid of a timing diagram, explain why the timing for transferring information to the data bus may be different for read cycles than for write cycles in a microcomputer.

14 Name the control signals required to control the flow of data between a 6502 microprocessor and its memory or peripheral devices, and, with the aid of a timing diagram, explain the behaviour of each signal.

15 Name the control signals required to control the flow of data between a Z80 microprocessor and its memory or peripheral devices, and, with the aid of a timing diagram, explain the behaviour of each signal.

16 Name the control signals required to control the flow of data between a 6800 microprocessor and its memory or peripheral devices, and, with the aid of a timing diagram, explain the behaviour of each signal.

4 Microprocessor instruction sets and machine code programs

A   MAIN POINTS CONCERNED WITH MICROPROCESSOR INSTRUCTION SETS AND MACHINE CODE PROGRAMS

1   The function of a microcomputer is to **receive data** from the outside world, **process the data** and **send results** back to the outside world. Within each micro-computer is a microprocessor which is capable of performing given tasks in response to certain binary inputs. These binary inputs are called '**instructions**', and a sequence of instructions which enable a microprocessor to perform a complete task is called a '**program**'. The range of instructions that a microprocessor recognises is called its '**instruction set**', and the nature of a particular instruction set depends very much on the internal construction (or 'architecture') of the microprocessor concerned.

2   A microprocessor receives instructions and data in the same form. Both are binary numbers, stored in memory and brought into the microprocessor by means of a system data bus. Instructions are distinguished from data by their order within a program and are fed into the instruction register of a microprocessor. Data are fed into one of the data registers or the arithmetic logic unit (ALU) of a micro-processor.

3   Data to be manipulated by a microprocessor are stored in one of its internal registers, or in external memory, and its location must be specified in each instruction. Many different ways exist for specifying data locations, and these are known as **addressing modes**. Some of the more commonly used addressing modes are as follows:

(i)   **implied** (or inherent);
(ii)   **immediate**;
(iii)   **relative**;
(iv)   **zero page** (or direct);
(v)   **absolute** (or extended);
(vi)   **indexed**; (see *Worked Problem 8*).
(vii)   **register direct**; and
(viii)   **register indirect** (see *Worked Problem 7*).

Simple programs (such as the addition of two numbers described in paragraph 6) may only use one form of addressing, but more complex programs may use many different types of addressing.

4   Most microprocessor instructions fall into one of the following three categories:
(i)   **data transfer** instructions,
(ii)   **arithmetic and logic** instructions, and
(iii)   **test and branch** instructions.

Examples of instructions from each of these three groups for typical microprocessors are given in *Tables 1, 2 and 3*.

Table 1 Typical data transfer instructions

Instruction to transfer data from:	Symbolic notation	Typical instructions for:		
		6502	Z80	6800
memory to accumulator	M → A	LDA	LD A, (nn)	LDAA
accumulator to memory	A → M	STA	LD (nn), A	STAA
memory to register	M → R	LDX	LD HL,(nn)	LDX
register to memory	R → M	STX	LD (nn),HL	STX
register to register	Rm → Rn	TXS	LD H,L	TXS

Note: '(nn)' is the contents of a memory location 'nn' in the range $\emptyset - 65536_{10}$ ($\emptyset - FFFF_{16}$)

Table 2(a) Typical arithmetic instructions

Instruction	Symbolic notation	Typical instructions for:		
		6502	Z80	6800
ADD (memory to accumulator)	A + M → A,C	–	ADD A,(HL)	ADD A
ADD WITH CARRY (memory to accumulator)	A + M + C → A,C	ADC	ADC A,(HL)	ADCA
SUBTRACT (memory from accumulator)	A − M → A,C	–	SUB (HL)	SUBA
SUBTRACT WITH CARRY (memory from accumulator)	A − M − C → A,C	SBC	SBC (HL)	SBCA
INCREMENT (memory)	M + 1 → M	INC	INC (HL)	INC
DECREMENT (memory)	M − 1 → M	DEC	DEC (HL)	DEC

Note: '(HL)' is the contents of a memory location defined by the HL register pair.

Table 2(b) Typical logic instructions

Instruction	Symbolic notation	Typical instructions for:		
		6502	Z80	6800
AND (memory with accumulator)	A · M → A	AND	AND (HL)	ANDA
OR (memory with accumulator)	A + M → A	ORA	OR (HL)	ORAA
EXCLUSIVE − OR (memory with accumulator)	A ⊕ M → A	EOR	XOR (HL)	EORA

Table 2(c) Typical shift and rotate instructions

Instruction	Symbolic notation	Typical instructions for:		
		6502	Z80	6800
LOGICAL SHIFT RIGHT (memory or accumulator)		LSR	SRL	LSR
ARITHMETIC SHIFT RIGHT (memory or accumulator)		—	SRA	ASR
ARITHMETIC SHIFT LEFT (memory or accumulator)		ASL	SLA	ASL
ROTATE RIGHT (memory or accumulator)		ROR	RR	ROR
ROTATE LEFT (memory or accumulator)		ROL	RL	ROL

Table 3 Test and branch instructions

Instruction	Symbolic notation	Typical instructions for:		
		6502	Z80	6800
COMPARE (memory with accumulator)	$A - M$	CMP	CP (HL)	CMPA
BIT	$A \cdot M$	BIT	*	BITA
JUMP/BRANCH (unconditional)	$nn \rightarrow PC$	JMP	JP/JR	JMP/BRA
JUMP/BRANCH (conditional)	$nn \rightarrow PC$ if condition is true, otherwise continue.	B − −	JP − − or JR − − depends upon condition imposed	B − −

Note: 'nn' is an address in the range $\emptyset$ to $65,536_{10}$ ($\emptyset - FFFF_{16}$)

*The Z80 microprocessor has a very much more comprehensive set of instructions for setting, resetting and testing bits — see Z80 instruction set, *Appendix B*

5 In order to create any program, a logical procedure must be adopted. A method frequently chosen consists of the following steps:
  (i) **Problem definition** The nature of a problem to which a solution is being sought must be clearly defined before programming can take place, otherwise the

resultant program may not perform its required task adequately. An eloquent program which solves the wrong problem is of no use at all. Some of the points which may need defining can be listed as follows:

(a) **state precisely what function the program is to perform;**
(b) **list the exact numbers and types of inputs and outputs required;**
(c) **state any constraints regarding speed of operation, accuracy; memory size or other physical factors; and**
(d) **state what action is to take place if operating errors occur.**

(ii) **Algorithms** An algorithm consists of a sequence of steps which define a method of solving a particular problem. It is possible for a problem to have more than one algorithm, i.e. there is more than one way of solving a particular problem. An algorithm (or lack of algorithm) may also indicate that a problem has no solution.

(iii) **Flowcharts** A flowchart is a graphical method of representing an algorithm. Standard symbols of the type illustrated in Chapter 1 are used in the preparation of flowcharts. Each step of an algorithm is represented by one or more of these symbols, suitably labelled with its function, and linked together in such a manner as to represent program flow (see examples of flow charts in the worked problems of this chapter).

Except in the case of very elementary problems, it is not normally possible to write a satisfactory program without first preparing a flowchart. A flowchart provides an overall view of the solution to a problem, thus minimising the possibility of the occurrence of logical errors and duplication. Flowcharts are also invaluable in program documentation, since they enable a program user to follow the logical thought processes of the programmer more clearly.

(iv) **Coding** This is the process of translating a flow chart into instructions which a microprocessor can execute. A microprocessor is only capable of interpreting instructions provided in binary form and known as **'machine code'**. This form of coding is rather tedious and error-prone for human operators, therefore machine code programs are usually written in either hexadecimal codes, or by using symbolic notation known as **'assembly language'**. *Table 4* shows an example of assembly language and its relationship to machine code. The main advantages of assembly language are:

(a) the use of instruction **'mnemonics'** makes a program more readable and avoids the necessity of remembering instruction opcodes:
(b) the use of **'symbolic addresses'** for memory locations avoids the necessity of keeping track of addresses and does not require address changes if instructions are added or deleted; and
(c) program writing is much faster and less error-prone.

Conversion from assembly language to machine code may be performed by hand — **'hand-assembly'** or by means of a computer program called an **'assembler'**.

(v) **Testing and debugging** Once a machine code program is written, it may be loaded into a microcomputer memory and tested for correct operation. Except for very simple routines, few programs perform as intended at the first attempt. Various errors may occur in the program-writing processes which lead to partial or total malfunctioning of a program. These errors are generally referred to as **'bugs'**, and the process of tracking down and eliminating bugs is called **'debugging'**.

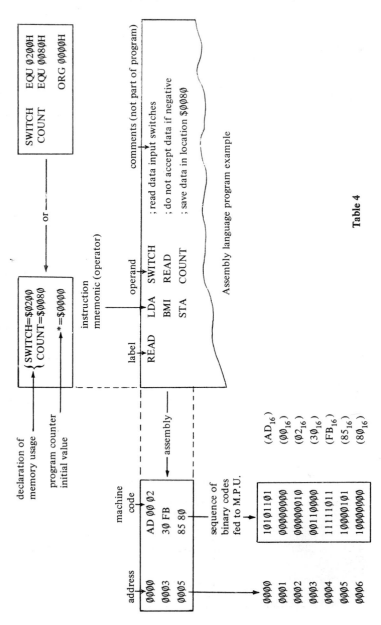

Table 4

Assembly language program example

declaration of memory usage

SWITCH	EQU 0200H
COUNT	EQU 0080H
	ORG 0000H

or

{ SWITCH = $0200
COUNT = $0080
* = $0000 }

program counter initial value

label	instruction mnemonic (operator)	operand	comments (not part of program)
READ	LDA	SWITCH	; read data input switches
	BMI	READ	; do not accept data if negative
	STA	COUNT	; save data in location $0080

assembly

machine code

address	
0000	AD 00 02	(AD₁₆)
0003	30 FB	
0005	85 80	

sequence of binary codes fed to M.P.U.

address		
0000	10101101	(AD₁₆)
0001	00000000	(00₁₆)
0002	00000010	(02₁₆)
0003	00110000	(30₁₆)
0004	11111011	(FB₁₆)
0005	10000101	(85₁₆)
0006	10000000	(80₁₆)

98

6 The following sample program demonstrates the use of the above procedures:

Problem: Perform an unsigned addition between two numbers located in memory at 0080_{16} and 0081_{16} and put the resulting sum into location 0082_{16}. Any carry resulting from this addition sets the carry flag but no further action is taken. No special requirements exist regarding speed of operation. Machine code for (a) 6502, (b) Z80 and (c) 6800 microprocessors is required.

Algorithm:
(i) clear carry flag (if necessary);
(ii) load the accumulator with the first number (location 0080_{16});
(iii) add on the second number (location 0081_{16});
(iv) store the sum in memory location 0082_{16}; and
(v) stop.

Flowchart: see *Figure 1*.

Figure 1

Program code:

(a) **6502**

Machine code		AUGEND ADDEND SUM *	=$0080 =$0081 =$0082 =$0020	Assembly language	
0020	18			CLC	; clear carry flag
0021	A5 80			LDA AUGEND	; get first number into accumulator
0023	65 81			ADC ADDEND	; add on second number
0025	85 82			STA SUM	; save the sum
0027	00			BRK	; program end

(b) **Z80** (Note, symbols in parenthesis refer to 'contents of a location' or 'contents of location pointed at by a particular register pair'.)

Machine code		AUGEND ADDEND SUM	EQU 0C80H EQU 0C81H EQU 0C82H ORG 0C90H	Assembly language	
0C90	21 81 0C			LD HL, ADDEND	; HL register pair points to addend
0C93	3A 80 0C			LD A, (AUGEND)	; get first number into accumulator
0C96	86			ADD A, (HL)	; add on second number
0C97	23			INC HL	; point HL to SUM location
0C98	77			LD (HL), A	; save the sum
0C99	76			HALT	; program end

(c) 6800

Machine Code		AUGEND ADDEND SUM	EQU $0080 EQU $0081 EQU $0082 ORG $0020	Assembly Language
0020	96 80		LDAA AUGEND	; get first number into accumulator A
0022	9B 81		ADDA ADDEND	; add on second number
0024	97 82		STAA SUM	; save the sum
0026	3F		SWI	; program end

Debugging: At this stage, a program may be checked for possible errors by constructing a 'trace table'. This is a table which shows the contents of each register and relevant memory location before and after the execution of each instruction in a program. A trace table for the above program for each of the three microprocessors used appears in *Table 5(a) to (c)*. An assumption is made that the two numbers added are 03_{16} and 05_{16} respectively.

7 Programs are usually loaded into memory by means of a keyboard, or, via a transfer medium such as paper tape, magnetic tape or floppy disc. Software is required to read data from these devices into memory, but until they are operational, no way exists of entering software into a microcomputer. This dilemma is resolved by having software already present in a microcomputer at switch-on time, i.e. some **firmware** is present. This firmware forms part of what is called the microcomputer '**operating system**' or '**monitor**'. Monitor programs may also contain aids for debugging programs such as setting breakpoints, single stepping, printing register contents and tabulating memory contents onto a monitor screen.

Table 5(a) 6502 trace table

Address bus	Instruction mnemonic	Acc	Memory locations			Flag register				Program counter
			0080	0081	0082	C	N	V	Z	
		XX	03	05	XX	X	X	X	X	0020
0020	CLC	XX	03	05	XX	0	X	X	X	0021
0021	LDA $80	03	03	05	XX	0	0	X	0	0023
0023	ADC $81	08	03	05	XX	0	0	0	0	0025
0025	STA $82	08	03	05	08	0	0	0	0	0027
0027	BRK									

Table 5(b) Z80 trace table

Address bus	Instruction mnemonic	A	HL	∅C8∅	∅C81	∅C82	C	S	V	Z	Program counter
		XX	XXXX	∅3	∅5	XX	X	X	X	X	∅C9∅
∅C9∅	LD HL, ∅C81H	XX	∅C81	∅3	∅5	XX	X	X	X	X	∅C93
∅C93	LD A (∅C8∅H)	∅3	∅C81	∅3	∅5	XX	X	X	X	X	∅C96
∅C96	ADD A, (HL)	∅8	∅C81	∅3	∅5	XX	∅	∅	∅	∅	∅C97
∅C97	INC HL	∅8	∅C82	∅3	∅5	XX	∅	∅	∅	∅	∅C98
∅C98	LD (HL), A	∅8	∅C82	∅3	∅5	∅8	∅	∅	∅	∅	∅C99
∅C99	HALT										

Table 5(c) 6800 trace table

Address bus	Instruction mnemonic	Acc A	∅∅8∅	∅∅81	∅∅82	C	N	V	Z	Program counter
		XX	∅3	∅5	XX	X	X	X	X	∅∅2∅
∅∅2∅	LDAA $8∅	∅3	∅3	∅5	XX	X	∅	∅	∅	∅∅22
∅∅22	ADDA $81	∅8	∅3	∅5	XX	∅	∅	∅	∅	∅∅24
∅∅24	STAA $82	∅8	∅3	∅5	∅8	∅	∅	∅	∅	∅∅26
∅∅26	SWI									

B   WORKED PROBLEMS ON MICROPROCESSOR INSTRUCTION SETS AND MACHINE CODE PROGRAMS

Problem 1   Explain the operation of the following addressing modes:
(i) immediate;   (iii) absolute (or extended);
(ii) zero − page (or direct)   (iv) implied (or inherent).

(i) In the **immediate** mode of addressing, the actual data to be processed forms part of the instruction. The data occupies a memory location immediately following the memory location in which an instruction operator is stored. For example, consider the instruction:

ADC#$∅8   (# signifies immediate mode).

Executing this instruction causes $∅8_{16}$ to be added to the original contents of the accumulator. This process is illustrated for typical microprocessors in

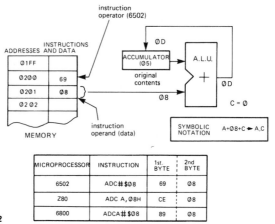

MICROPROCESSOR	INSTRUCTION	1st. BYTE	2nd BYTE
6502	ADC#$08	69	08
Z80	ADC A,08H	CE	08
6800	ADCA#$08	89	08

Figure 2

Figure 2. Immediate addressing is efficient in memory usage, and is fast in operation, but is only suitable for use involving known constants.

(ii) With **zero − page** (or direct) addressing, the second byte of an instruction contains the lower eight bits of the address of the operand. This permits up to 256 different locations to be referenced, usually in the range 0000_{16} to $00FF_{16}$ (hence the term 'zero − page'), although some microprocessors (e.g. Motorola 6809) allow a user to specify the range. As an example of zero-page addressing, consider the following instruction:

ADC $0025 .

Executing this instruction causes the contents of memory location 0025_{16} to be added to the original contents of the accumulator. This process is illustrated for typical microprocessors in *Figure 3*. This form of addressing is efficient in memory usage, and is fast in operation since a microprocessor has only to fetch part of the address of the operand.

Note: A Z80 microprocessor does not have this form of addressing, but due to its register oriented construction, many operations may be performed with single byte instructions.

(iii) With **absolute** (or extended) addressing, the full address of the operand is specified. Therefore, when using this addressing mode, instructions are three bytes long. For example, consider the following instruction:

ADC $0250

Executing this instruction causes the contents of memory location 0250_{16} to be added to the original contents of the accumulator. This process is illustrated for typical microprocessors in *Figure 4*.

Note: Both the 6502 and Z80 microprocessors use a system known as 'low order first' in which the lower half of the absolute address is stored as the second byte of an instruction, and the upper half is stored as the third byte.

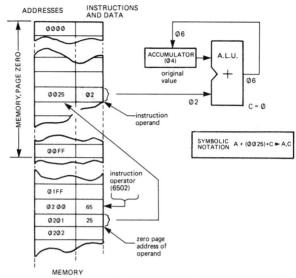

MICROPROCESSOR	INSTRUCTION	1st BYTE	2nd BYTE
6502	ADC $0025	65	25
6800	ADCA $0025	99	25

Figure 3

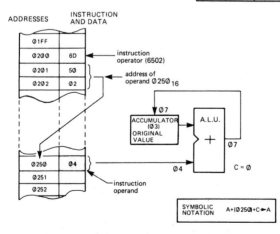

MICROPROCESSOR	INSTRUCTION	1st. BYTE	2nd. BYTE	3rd. BYTE
6502	ADC $0250	6D	50	02
6800	ADCA $0250	B9	02	50

Figure 4

(iv) **Implied** (or inherent) addressing is used with instructions which do not require an operand, or the address of an operand, to be specified. These are therefore single byte instructions in which the source and destination addresses for data are implied in the instruction. Examples of instructions using this form of addressing are as follows:

    (a) **CLC** (clear carry flag);   (b) **CLRA** (clear accumulator A);

    (c) **INCA** (increment accumulator A);

    (d) **ASLA** (arithmetic shift left accumulator);

    (e) **TAX** (transfer accumulator contents to index register X);

    (f) **NOP** (no operation − do nothing except increment the program counter); and (g) **DAA** (decimal adjust the accumulator).

Problem 2 With reference to '**relative addressing**', explain:
(i)   what is meant by relative addressing;
(ii)   which types of instruction use this form of addressing;
(iii) how relative offsets are calculated; and
(iv) the main advantage of this type of addressing.

(i) This is a form of addressing in which the second byte of an instruction is treated as a **two's complement displacement** (or offset), added to the current value of the program counter to calculate an effective address. Thus the effective address for an instruction is specified relative to the program counter.

(ii) Relative addressing is used in a large number of microprocessors for **jump** and **branch** instructions.

(iii) When relative addressing is used for branch instructions, the second byte of an instruction must be of such a value as to enable the next instruction to be located if the branch is taken. The address of the next instruction after branching is called the '**effective address**', and may be expressed as:

> **Effective address = program counter + offset**

or   > **Offset = effective address − program counter**

The offset is an 8-bit two's complement number, and this allows **branching forward** $+127_{10}$ places, and **backwards** -128_{10} places relative to the program counter value. A program counter is incremented by one before each instruction or data fetch, therefore when a branch offset is fetched, the program counter is already advanced to the next location. This fact must be taken into account when calculating an offset.

Consider, for example, the situation where a branch instruction is located in memory at 0025_{16} (operator) and 0026_{16} (offset), and, if taken, a branch to location 0039_{16} is desired. This is illustrated in *Figure 5* and is called a '**forward branch**'.

Consider as a second example, a branch instruction located in memory at 0050_{16} (operator) and 0051_{16} (offset), and, if taken, a branch to location 0025_{16}. This is illustrated in *Figure 6* and is an example of a '**backward branch**'. From *Figures 5 and 6* it can be seen that the program counter is

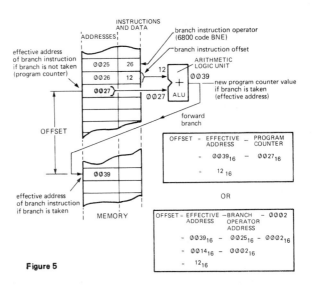

Figure 5

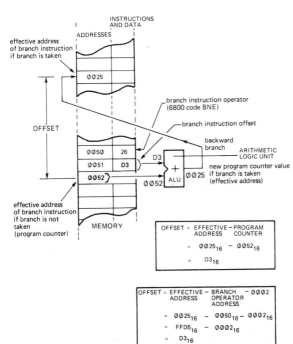

Figure 6

always two places ahead of the location of the branch operator. Therefore the value of an offset is sometimes stated as:

> Offset = effective address − address of branch operator − 0002

(iv) The main advantage of using this form of addressing is that programs may be moved to new locations in memory without the need to change any of the instructions, i.e. programs which use relative addressing are 'relocatable'.

A second advantage is that relative addressing requires two bytes for each instruction rather than the three bytes required for absolute addressing. Therefore less memory space is required and programs may run at a faster rate due to there being fewer data fetch cycles.

Problem 3 Explain the operation of '**Compare**' and '**Bit**' instructions and show how they differ from '**Subtract**' and '**AND**' instructions.

'Compare' and 'Bit' are both 'flag − setting' instructions and behave in the following manner:

A '**compare**' instruction causes the contents of a specified memory location to be subtracted from the contents of the accumulator (or other specified register), and causes appropriate flags in the status register to be set or reset according to the result of the subtraction. However, a compare instruction does **not** put the result of its subtraction back into the accumulator. This means that, since the accumulator contents are undisturbed, repeated comparisons may be made with the accumulator until a required condition is met (see *Figure 7*).

A '**bit**' instruction is the logical equivalent of a '**compare**' instruction. The contents of a specified memory location are logically ANDed with the contents of the accumulator and the appropriate flags in the status register are set according to the result of this operation. Unlike an 'AND' operation, the result of a 'Bit' operation is **not** put back into the accumulator. Thus individual bits of specified memory locations may be tested by a process of repeated ANDing with a mask which is stored in the accumulator. This process is illustrated in *Figure 8*.

Problem 4 Describe the operation and uses of the following microprocessor instructions:
(i) **Logical shift**; (ii) **Arithmetic shift**; and (iii) **Rotate**.

These instructions are used in a wide variety of processes which involve manipulation of the location of bits in a register or memory location. Examples of the uses of these instructions are:
(a) **serial to parallel conversion of data**;
(b) **parallel to serial conversion of data**;
(c) **arithmetical multiplication routines**;
(d) **arithmetical division routines**;
(e) **packing** and **unpacking of data** (combining or separating two 4-bit words); and

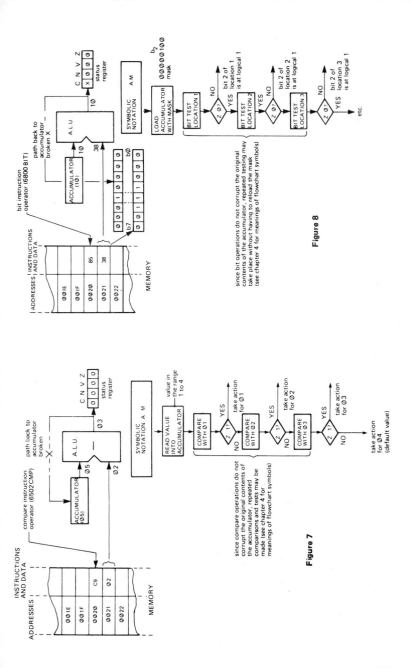

Figure 8

since bit operations do not corrupt the original contents of the accumulator, repeated testing may take place without having to reload the mask (see chapter 4 for meanings of flowchart symbols)

Figure 7

since compare operations do not corrupt the original contents of the accumulator, repeated comparisons and tests may be made (see chapter 4 for meanings of flowchart symbols)

107

(f) **matching bit patterns.**

(i) A **logical shift** operation causes each bit in a register or memory location to be moved into the position previously occupied by its adjacent bit. A logical $\emptyset$ is moved into the empty position and bit $\emptyset$ or bit 7 (depending upon the direction of the shift) is moved into the carry flag. This process is illustrated in *Figure 9*.

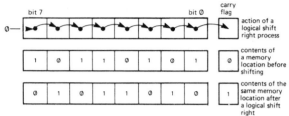

Figure 9 Logical shift right

(ii) An **arithmetic shift** left behaves in a similar manner to a logical shift except that the **sign bit is retained**. An arithmetic shift right is illustrated in *Figure 10*.

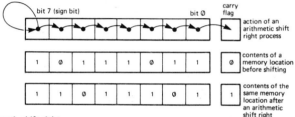

Figure 10 Arithmetic shift right

(iii) There are two possible forms of rotate instruction: (a) Rotate without carry, and (b) Rotate with carry.

The behaviour of rotate instructions is similar to that of shift instructions, except that bits shifted out from one end of a register or memory location (which are lost with shift operations), are shifted back in at the opposite end. This means, in effect, bit $\emptyset$ of a register or memory location is connected back to bit 7 during execution of rotate instructions, and via the carry flag to bit 7 during execution of rotate with carry instructions. This is illustrated in *Figures 11(a) and (b)*.

Problem 5 With reference to the extracts from the instruction sets for the 6502 and 6800 microprocessors, illusttated in *Figure 12*, explain the meaning of the columns labelled A to E .

Ⓐ This column is reserved for **instruction mnemonics**. Mnemonics, in this context, are symbolic names or abbreviations which suggest the function of the instruction. For example, ADC — ADd with Carry. Program writing is much easier when using mnemonics, since each instruction may be defined by only three or four letters.

108

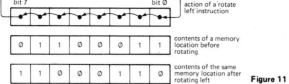

action of a 'rotate left' instruction

contents of a memory location before rotating

contents of the same memory location after rotating left

Figure 11 (a) Rotate left

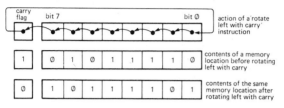

action of a 'rotate left with carry' instruction

contents of a memory location before rotating left with carry

contents of the same memory location after rotating left with carry

Figure 11 (b) Rotate left with carry

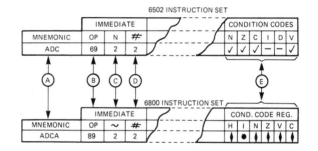

6502 INSTRUCTION SET

	IMMEDIATE			CONDITION CODES					
MNEMONIC	OP	N	#	N	Z	C	I	D	V
ADC	69	2	2	√	√	√	—	—	√

6800 INSTRUCTION SET

	IMMEDIATE			COND. CODE REG.					
MNEMONIC	OP	~	#	H	I	N	Z	V	C
ADCA	89	2	2	↕	●	↕	↕	↕	↕

Figure 12

(B) This column is reserved for the instruction **operation codes** or **'opcodes'**. Operation codes form the operator part of an instruction and inform a microprocessor of what action it is to take. A microprocessor accepts operation codes in binary form, but for convenience, these are listed in an instruction set in hexadecimal form.

(C) This column is reserved to indicate the **number of cycles** necessary to execute an instruction. As shown in *Worked Problem 7*, Chapter 3, an instruction consists of a number of small steps called 'microinstructions'. Each of these steps is carried out in synchronism with pulses from a system timing clock (which is frequently built into the microprocessor itself). Therefore, the number of clock pulses (or cycles) required for each instruction is determined by the complexity of the instruction being considered. If the time for each clock cycle is known, the time for executing an instruction may be calculated by multiplying this time by the number of cycles in column C. This enables the complete running time for a program to be calculated. A typical clock frequency is 1 MHz, in which case each clock cycle has a duration of 1 μs.

(D) This column indicates the **number of bytes** required for each instruction, and hence enables the total number of memory locations required for a program to be determined. Most instructions consist of one, two or three bytes depending upon the addressing mode specified.

(E) This group of columns indicates the effect that executing an instruction has on the **flags** of a microprocessor **status register** (condition codes register). Columns which are marked with '√' or ' ↕ ' indicate that these particular flags are set or reset by an instruction according to the result of its operation. Columns which are marked with '−' or '•' indicate that these flags remain unaltered after the execution of an instruction. Other conditions which may be found in these columns are:
(a) R or ∅; flag reset always on execution of this instruction, and
(b) S or 1; flag set always on execution of this instruction.

Problem 6   Explain the operation of the following instructions:
(i) **branch if minus**;   (iv) **branch if not zero**;
(ii) **branch if plus**;   (v) **branch if carry set**; and
(iii) **branch if zero**;   (vi) **branch if carry clear**.

(i) If the result of an arithmetic or logic operation (or 'load' operation in some microprocessors) is negative, the N-flag in a microprocessor status register is set (N = 1). When a 'branch if minus' instruction is encountered in a program, the branch is taken if N = 1, but ignored if N = ∅. This is illustrated in *Figure 13*.

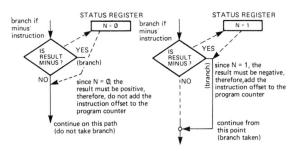

Figure 13

(ii) This instruction is the **complementary branch** to 'branch if minus'. The branch is taken if N = ∅, but is ignored if N = 1. This is illustrated in *Figure 14*.

(iii) If the result of an arithmetic or logic operation (or 'load' operation in some microprocessors) is zero, the Z-flag in a microprocessor status register is set, (Z = 1). When a 'branch if zero' instruction is encountered in a program, the branch is taken if Z = 1, but ignored if Z = ∅. This is illustrated in *Figure 15*.

(iv) This instruction is the **complementary branch** to 'branch if zero'. The branch is taken if Z = ∅, but is ignored if Z = 1. This is illustrated by *Figure 16*.

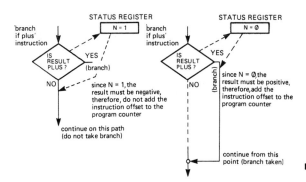

Figure 14

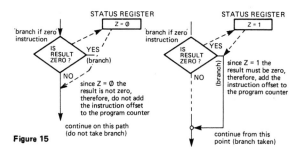

Figure 15

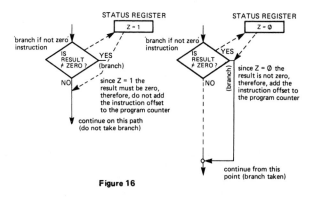

Figure 16

(v) Arithmetic operations, shifts and rotates all control the state of the carry flag in a microprocessor. When a 'branch if carry set' instruction is encountered in a program, the branch is taken if $C = 1$, but ignored if $C = \emptyset$. This is illustrated in *Figure 17*.

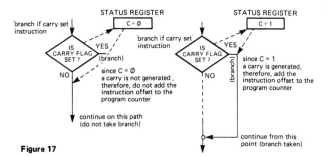

Figure 17

(vi) This instruction is the **complementary branch** to 'branch if carry set'. The branch is taken if C = ∅, but is ignored if C = 1. This is illustrated by *Figure 18*.

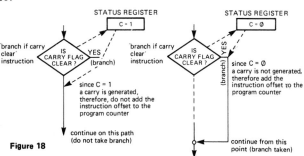

Figure 18

Problem 7   Explain the function of a **memory address register**. Using Z80 microprocessors as an example, show how memory address register may be used to transfer data from memory location $\emptyset C9\emptyset_{16}$ to $\emptyset E5\emptyset_{16}$.

Some microprocessors access data by specifying its memory address as part of an instruction. Such microprocessors are said to be '**memory oriented**'. Other microprocessors access data via memory pointer registers called '**store address registers**' or '**memory address registers**'. Such microprocessors are said to be '**register oriented**'.

A Z80 microprocessor has three sixteen-bit memory address registers, each formed from a pair of largely independent eight-bit registers (see *Figure 19*).

In order to transfer data from memory location $\emptyset C9\emptyset_{16}$ to location $\emptyset E5\emptyset_{16}$, an LDI or LDD instruction may be used, as shown in *Figure 20*.

Note that this operation does not change the contents of memory location $\emptyset C9\emptyset_{16}$, but merely stores a copy of its contents in $\emptyset E5\emptyset_{16}$. Therefore, after the data transfer takes place, $\emptyset C9\emptyset_{16}$ and $\emptyset E5\emptyset_{16}$ contain identical data. Also, the data transfer instruction (LDI or LDD) changes the addresses stored in HL and DE.

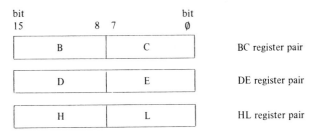

B	C	BC register pair
D	E	DE register pair
H	L	HL register pair

bit 15   8   7   bit 0

Figure 19

There is also a duplicate set of these registers available, but only one set may be in use at any given time.

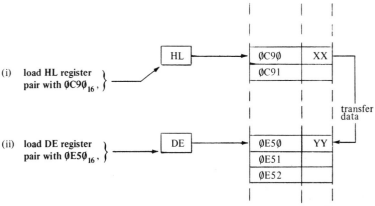

(i)   **load HL register pair with** $0C90_{16}$,

(ii)   **load DE register pair with** $0E50_{16}$,

(iii)   **transfer the data** pointed to by HL to the memory location to which DE points.

Figure 20

Problem 8   Explain the function of an index register in a microprocessor.

An index register is an 8-bit or 16-bit register within a microprocessor to which special instructions relate. An index register may be used for many general purpose applications such as temporary data storage, or as a binary counter, but its main application is for a method of accessing data known as 'indexed addressing'. When using this form of addressing, an instruction may access many different memory locations (such as data tables) rather than specifying a fixed memory location. The basic form of indexed addressing is to specify the actual memory location accessed (the effective address) as follows:

$$\begin{matrix} \text{Effective} \\ \text{address} \end{matrix} = \begin{matrix} \text{Base} \\ \text{address} \end{matrix} + \begin{matrix} \text{Address} \\ \text{displacement} \end{matrix} \text{(modifier)}$$

In some microprocessors, an instruction operand contains the **base address** and the index register contains a **displacement** (or modifier). In other microprocessors

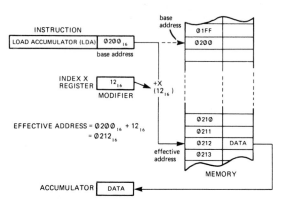

Figure 21 (a) 6502 indexed addressing

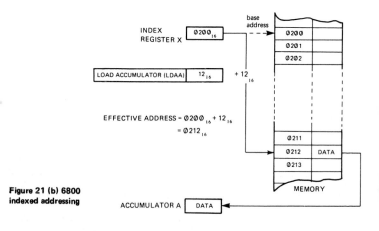

Figure 21 (b) 6800 indexed addressing

the index register holds the **base address** and the instruction operand contains a **displacement** (or modifier). These two types of indexed addressing are illustrated in *Figure 21(a) and (b)*.

Problem 9 Write a program (including algorithm and flowchart), using 6502 machine code, which exchanges data in memory locations $\emptyset\emptyset8\emptyset_{16}$ and $\emptyset\emptyset81_{16}$.

The program may be located in any convenient area of memory.

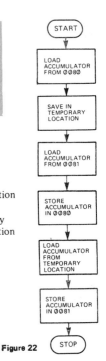

Algorithm:

(i) load the accumulator with the contents of memory location $\emptyset\emptyset8\emptyset_{16}$ and store this data in a temporary storage location,

(ii) load the accumulator with the contents of memory location $\emptyset\emptyset81_{16}$ and store this data in memory location $\emptyset\emptyset8\emptyset_{16}$,

(iii) load the accumulator with the data in the temporary storage location and store this data in memory location $\emptyset\emptyset81_{16}$, and

(iv) stop.

Flowchart: see *Figure 22*.

Figure 22

Program code:

Machine code			Assembly language
	DATA 1	=$\$\emptyset\emptyset8\emptyset$	
	DATA 2	=$\$\emptyset\emptyset81$	
	TEMP	=$\$\emptyset\emptyset82$	
	*	=$\$\emptyset\emptyset2\emptyset$	
$\emptyset\emptyset2\emptyset$ A5 8$\emptyset$		LDA DATA 1	; load accumulator from $\$\emptyset\emptyset8\emptyset$
$\emptyset\emptyset22$ 85 82		STA TEMP	; save in temporary location
$\emptyset\emptyset24$ A5 81		LDA DATA 2	; load accumulator from $\$\emptyset\emptyset81$
$\emptyset\emptyset26$ 85 8$\emptyset$		STA DATA 1	; store accumulator in $\$\emptyset\emptyset8\emptyset$
$\emptyset\emptyset28$ A5 82		LDA TEMP	; get data from temporary
			; location
$\emptyset\emptyset2A$ 85 81		STA DATA 2	; store accumulator in $\$\emptyset\emptyset81$
$\emptyset\emptyset2C$ $\emptyset\emptyset$		BRK	; program end

Problem 10 Construct a trace table for the program written in *Problem 9*.

For the purpose of constructing this trace table it is assumed that memory locations $\emptyset\emptyset8\emptyset_{16}$ and $\emptyset\emptyset81_{16}$ contain $\emptyset3_{16}$ and $\emptyset5_{16}$ initially.

Address bus	Instruction mnemonic	Acc	Memory locations			Flag register				Program counter
			ØØ8Ø	ØØ81	ØØ82	C	N	V	Z	
		XX	Ø3	Ø5	XX	X	X	X	X	ØØ2Ø
ØØ2Ø	LDA $8Ø	Ø3	Ø3	Ø5	XX	X	Ø	C	Ø	ØØ22
ØØ22	STA $82	Ø3	Ø3	Ø5	Ø3	X	Ø	X	Ø	ØØ24
ØØ24	LDA $81	Ø5	Ø3	Ø5	Ø3	X	Ø	X	Ø	ØØ26
ØØ26	STA $8Ø	Ø5	Ø5	Ø5	Ø3	X	Ø	X	Ø	ØØ28
ØØ28	LDA $82	Ø3	Ø5	Ø5	Ø3	X	Ø	X	Ø	ØØ2A
ØØ2A	STA $81	Ø3	Ø5	Ø3	Ø3	X	Ø	X	Ø	ØØ2C
ØØ2C	BRK									

Problem 11   Write a program (including algorithm and flowchart), using 6502 code, to subtract hex. data in memory location ØØ81$_{16}$ from hex data in memory location ØØ8Ø$_{16}$ and store the difference in memory location ØØ82$_{16}$. The program may be located in any convenient area of memory.

Algorithm:
(i)   set the carry flag;
(ii)   clear the decimal mode flag;
(iii)   load the accumulator again with data from memory location ØØ8Ø$_{16}$,
(iv)   logical AND the accumulator with ØF$_{16}$ to mask off the most significant four bits, and store the result in memory location ØØ81$_{16}$, and
(v)   store the difference in memory location ØØ82$_{16}$, and
(vi)   stop.

Flowchart: see *Figure 23*.

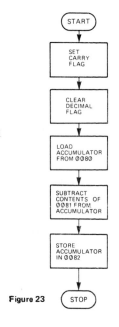

Figure 23

Program code:

Machine code		DATA 1	=$0080	Assembly language
		DATA 2	=$0081	
		DIFF	=$0082	
		*	=$0020	
0020	38		SEC	; set carry flag
0021	D8		CLD	; clear decimal mode
0022	A5 80		LDA DATA 1	; get minuend
0024	E5 81		SBC DATA 2	; subtract subtrahend
0026	85 82		STA DIFF	; store difference in location ; $0082
0028	00		BRK	; program end

Problem 12 Construct a trace table for the program written in *Problem 11*.

For the purpose of constructing this trace table it is assumed that memory locations 0080_{16} and 0081_{16} contain 09_{16} and 06_{16} initially.

Address bus	Instruction mnemonic	Acc	Memory locations			Flag register				Program counter
			0080	0081	0082	C	N	V	Z	
		XX	09	06	XX	X	X	X	X	0020
0020	SEC	XX	09	06	XX	1	X	X	X	0021
0021	CLD	XX	09	06	XX	1	X	X	X	0022
0022	LDA $80	09	09	06	XX	1	X	X	X	0024
0024	SBC $81	03	09	06	XX	1	0	0	0	0026
0026	STA $82	03	09	06	03	1	X	X	X	0028
0028	BRK									

Problem 13 Write a program (including algorithm and flowchart), using 6502 code, which converts a hexadecimal number in the range 00_{16} to $0F_{16}$ into its decimal equivalent and stores it in memory location 0081_{16}. Assume that the hexadecimal number to be converted is located in memory at 0080_{16} and that the program may be located in any convenient area of memory.

Algorithm:

(i) load the accumulator with the hex. number to be converted,

(ii) compare the value in the accumulator with $0A_{16}$;

(iii) if the N-flag is set (N = 1), the number is in the range 00_{16} to 09_{16} and no adjustment is necessary;

(iv) if the N-flag is clear (N = 0), the number is in the range $0A_{16}$ to $0F_{16}$ and 06_{16} must be added;

(v) store the decimal equivalent of the number in memory location 0081_{16}; and

(vi) stop.

Flowchart: see *Figure 24*.

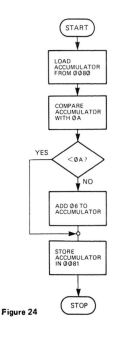

Figure 24

Program code:

Machine code			HEXNUM	=$0080	Assembly language
			DECNUM	=$0081	
			*	=$0020	
0020	A5	80		LDA HEXNUM	; load accumulator with hex ; number
0022	C9	0A		CMP #$0A	; compare with $0A
0024	30	03		BMI SAVE	; do not adjust number
0026	18			CLC	
0027	69	06		ADC #$06	; convert to decimal
0029	85	81	SAVE	STA DECNUM	; store decimal equivalent in ; $0081
002B	00			BRK	; program end

Problem 14 Memory location 0080_{16} contains a hexadecimal number in the range 00_{16} to 50_{16}. Write a program (including algorithm and flowchart), using 6502 code which multiplies this number by 5_{16} and stores the product in memory location 0081_{16}. The program may be located in any convenient area of memory.

Figure 25

Algorithm:

(i) load the accumulator with the contents of memory location 0080_{16};

(ii) shift the contents of the accumulator two places to the left (equivalent to multiplying by 4);

(iii) add the contents of memory location 0080_{16} to the accumulator (with (ii), is equivalent to multiplying by 5);

(iv) store the contents of the accumulator in memory location 0081_{16}, and

(v) stop

Flowchart: see *Figure 25*

Program code:

Machine code			HEXNUM=$0080 PROD=$0081 *=$0020	Assembly language
0020	A5	80	LDA HEXNUM	; load accumulator with number
0022	0A		ASL A	
0023	0A		ASL A	; shift left twice
0024	18		CLC	; prepare for addition
0025	65	80	ADC HEXNUM	; add on hex. number
0027	85	81	STA PROD	; store the product in $0081
0029	00		BRK	; program end

Problem 15 Write a program (including algorithm and flowchart), using 6502 code, which checks a data word in memory location 0080_{16} and stores a value (flag) in memory location 0081_{16} with the following meaning:

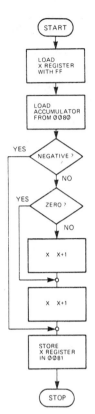

Algorithm:

(i) load index register X with FF_{16};

(ii) load the accumulator with the data word from memory location 0080_{16};

(iii) if the data word is negative, store the contents of index register X in memory location 0081_{16} and stop;

(iv) if the data word is zero, increment index register by 1 and store its contents in memory location 0081_{16} and stop;

(v) if the data word is positive, increment index register by 2 and store its contents in memory location 0081_{16} and stop.

Flowchart: see *Figure 26*.

Figure 26

Program code:

Machine code			WORD	=$0080	Assembly language
			FLAG	=$0081	
			*	=$0020	
0020	A2	FF		LDX #$FF	; load index register X with $FF
0022	A5	80		LDA WORD	; load accumulator with data
					; word
0024	30	04		BMI MINUS	; test for negative data word
0026	F0	01		BEQ ZERO	; test for zero data word
0028	E8			INX	; must be positive by default
0029	E8		ZERO	INX	; increment X as appropriate
002A	86	81	MINUS	STX FLAG	; store X in memory location
					; $0081
002C	00			BRK	; program end

120

Problem 16   Two electrical sensors in a control system provide two 4 bit words which are stored together in a single memory location (at 0080_{16}) as a single 8 bit word as illustrated in *Figure 27*. Write a program (including algorithm and flowchart), using 6502 code, to separate out (unpack) the two 4 bit words and store them in two separate memory locations, 0081_{16} and 0082_{16} such that they occupy the lower four bits in each of their respective memory locations.

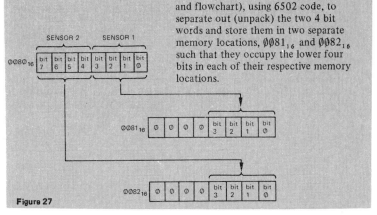

Figure 27

Algorithm:

(i)   load the accumulator with data from memory location 0080_{16},

(ii) shift the accumulator contents right 4 times and store the result in memory location 0082_{16},

(iii) load the accumulator again with data from memory location 0080_{16},

(iv) logical AND the accumulator with $0F_{16}$ to mask off the most significant four bits, and store the result in memory location 0081_{16}, and

(v)   stop.

Flowchart: see *Figure 28*.

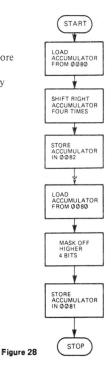

Figure 28

121

Program code:

Machine code	DATA =$0080 SENS1=$0081 SENS2=$0082 *=$0020	Assembly language	
0020 A5 80		LDA DATA	; load accumulator with data
0022 4A		LSR A	
0023 4A		LSR A	; shift higher 4 bits into lower
0024 4A		LSR A	; 4 bit positions
0025 4A		LSR A	
0026 85 82		STA SENS2	; store sensor 2 data in $0082
0028 A5 80		LDA DATA	; reload accumulator with data
002A 69 0F		AND #$0F	; mask off higher 4 bits
002C 85 81		STA SENS1	; store sensor 1 data in $0081
002E 00		BRK	; program end

Problem 17 Write a program (including algorithm and flowchart) using Z80 code which exchanges data in memory locations $0C80_{16}$ and $0C81_{16}$. The program may be located in any convenient area of memory.

Algorithm:
(i)   load register A (accumulator) with the contents of memory location $0C80_{16}$;
(ii) load register B with the contents of memory location $0C81_{16}$;
(iii) store the contents of register B in memory location $0C80_{16}$;
(iv) store the contents of register A in memory location $0C81_{16}$; and
(v)   stop.

Flowchart: see *Figure 29*.

Program code:

Machine code	DATA1 EQU 0C80H DATA2 EQU 0C81H ORG 0C90H	Assembly language
0C90 21 80 0C	LD HL,DATA1	; point HL register to DATA1
0C93 7E	LD A,(HL)	; load A with contents of DATA1
0C94 23	INC HL	; increment HL register by 1
0C95 46	LD B,(HL)	; load B with contents of DATA2
0C96 77	LD (HL),A	; store contents of A in 0C81H
0C97 2B	DEC HL	; decrement HL register by 1
0C98 70	LD (HL),B	; store contents of B in 0C80H
0C99 76	HALT	; program end

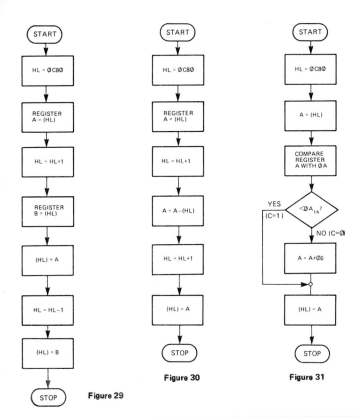

Figure 29

Figure 30

Figure 31

Problem 18 Construct a trace table for the program written in *Problem 17*.

Address bus	Instruction mnemonic	Registers			Memory locations		Flag register				Program counter
		A	B	HL	ØC8Ø	ØC81	C	S	V	Z	
		XX	XX	XXXX	Ø3	Ø5	X	X	X	X	ØC9Ø
ØC9Ø	LD HL, ØC8ØH	XX	XX	ØC8Ø	Ø3	Ø5	X	X	X	X	ØC93
ØC93	LD A, (HL)	Ø3	XX	ØC8Ø	Ø3	Ø5	X	X	X	X	ØC94
ØC94	INC HL	Ø3	XX	ØC81	Ø3	Ø5	X	X	X	X	ØC95
ØC95	LD B, (HL)	Ø3	Ø5	ØC81	Ø3	Ø5	X	X	X	X	ØC96
ØC96	LD (HL), A	Ø3	Ø5	ØC81	Ø3	Ø3	X	X	X	X	ØC97
ØC97	DEC HL	Ø3	Ø5	ØC8Ø	Ø3	Ø3	X	X	X	X	ØC98
ØC98	LD (HL), B	Ø3	Ø5	ØC8Ø	Ø5	Ø3	X	X	X	X	ØC99
ØC99	HALT										

123

Algorithm:

(i)   load register A (accumulator) with the contents of memory location $\emptyset C8\emptyset_{16}$;
(ii) subtract the contents of memory location $\emptyset C81_{16}$ from register A;
(iii) store the contents of register A in memory location $\emptyset C82_{16}$; and
(iv) stop.

Flowchart: see *Figure 30*.

Program code:

Machine code		DATA1	EQU	ØC80H	Assembly language
		DATA2	EQU	ØC81H	
		DIFF	EQU	ØC82H	
			ORG	ØC90H	
ØC90	21 80 ØC		LD HL,DATA1		; point HL register pair to ØC80H
ØC93	7E		LD A,(HL)		; load A with contents of DATA1
ØC94	23		INC HL		; increment HL by 1
ØC95	96		SUB (HL)		; subtract contents of DATA2
					; from A
ØC96	23		INC HL		; increment HL by 1
ØC97	77		LD (HL),A		; store contents of A in ØC82H
ØC98	76		HALT		; program end

For the purpose of constructing this trace table it is assumed that memory locations $\emptyset C8\emptyset_{16}$ and $\emptyset C81_{16}$ contain $\emptyset 9_{16}$ and $\emptyset 6_{16}$ initially.

Address bus	Instruction mnemonic	Registers		Memory locations			Flag register				Program counter
		A	HL	ØC8Ø	ØC81	ØC82	C	S	V	Z	
		XX	XXXX	Ø9	Ø6	XX	X	X	X	X	ØC9Ø
ØC9Ø	LD HL, ØC8Ø	XX	ØC8Ø	Ø9	Ø6	XX	X	X	X	X	ØC93
ØC93	LD A, (HL)	Ø9	ØC8Ø	Ø9	Ø6	XX	X	X	X	X	ØC94
ØC94	INC HL	Ø9	ØC81	Ø9	Ø6	XX	X	X	X	X	ØC95
ØC95	SUB (HL)	Ø3	ØC81	Ø9	Ø6	XX	Ø	Ø	Ø	Ø	ØC96
ØC96	INC HL	Ø3	ØC82	Ø9	Ø6	XX	Ø	Ø	Ø	Ø	ØC97
ØC97	LD (HL), A	Ø3	ØC82	Ø9	Ø6	Ø3	Ø	Ø	Ø	Ø	ØC98
ØC98	HALT										

Algorithm:

(i)   load register A (accumulator) with the hex. number to be converted,

(ii) compare the value in register A with $0A_{16}$,

(iii) if the C-flag is set (C = 1), the number is in the range 00_{16} to 09_{16} and no adjustment is necessary,

(iv) if the C-flag is clear (C = 0), the number is in the range $0A_{16}$ to $0F_{16}$, and 06_{16} must be added;

(v)   store the decimal equivalent of the number in memory location $0C81_{16}$; and

(vi) stop.

Flowchart: see *Figure 31*.

Program code:

Machine code		HEXNUM	EQU 0C80H	Assembly language
		DECNUM	EQU 0C81H	
			ORG 0C90H	
0C90	21 80 0C		LD HL, HEXNUM	; point HL register to 0C80H
0C93	7E		LD A, (HL)	; load A register with hex. ; number
0C94	FE 0A		CP 0AH	; compare with 0AH
0C96	38 02		JR C, SAVE	; do not adjust number
0C98	C6 06		ADD A, 06H	; convert to decimal
0C9A	23	SAVE	INC HL	; increment HL by 1
0C9B	77		LD (HL),A	; store decimal equivalent in ; 0C81H
0C9C	76		HALT	; program end

Algorithm:

(i)   load register A (accumulator) with the contents of memory location $0C80_{16}$.

(ii) shift the contents of register A two places to the left (equivalent to multiplying by 4),

(iii) add the contents of location $0C80_{16}$ to register A (with (ii); is equivalent to multiplying by 5),

(iv) store the contents of register A in memory location $0C81_{16}$; and

(v)   stop.

Flowchart: see *Figure 32*.

Program code:

Machine code	HEXNUM	EQU 0C80H	Assembly language	
	PROD	EQU 0C81H		
		ORG 0C90H		
0C90 21 80 0C		LD HL,HEXNUM	; point HL in register to 0C80H	
0C93 7E		LD A,(HL)	; load register A with hex. number	
0C94 87		ADD A,A	;	
0C95 87		ADD A,A	; equivalent to shift left A twice	
0C96 86		ADD A,(HL)	; add on hex. number	
0C97 23		INC HL	; increment register HL by 1	
0C98 77		LD (HL),A	; store the product in 0C81H	
0C99 76		HALT	; program end	

Problem 23 Write a program (including algorithm and flowchart), using Z80 code, which checks a data word in memory location $0C80_{16}$, and stores a value (flag) in memory location $0C81_{16}$ with the following meaning:
(i) FF_{16} if the data word is negative;
(ii) 00_{16} if the data word is zero; or
(iii) 01_{16} if the data word is positive.
The program may be located in any convenient area of memory.

Algorithm:
(i) load register B with FF_{16};
(ii) load register A (accumulator) with the contents of memory location $0C80_{16}$;
(iii) if the data word is negative, store the contents of register B in memory location $0C81_{16}$, and stop;
(iv) if the data word is zero, increment register B by 1, and store its contents in memory location $0C81_{16}$, and stop;
(v) if the data word is positive, increment register B by 2 and store its contents in memory location $0C81_{16}$, and stop.

Flowchart: see *Figure 33*.

Program code:

Machine code	WORD	EQU 0C80H	Assembly language
	FLAG	EQU 0C81H	
		ORG 0C90H	
0C90 21 80 0C		LD HL, WORD	; point HL register pair to 0C80H
0C93 06 FF		LD B,FFH	; load register B with FFH
0C95 7E		LD A,(HL)	; load register A with data word
0C96 F6 00		OR 00H	; set flags
0C98 FA 9F 0C		JP N,MINUS	; test for negative data word
0C9B 28 01		JR Z,ZERO	; test for zero data word
0C9D 04		INC B	; must be positive by default
0C9E 04	ZERO	INC B	; increment B as appropriate
0C9F 23	MINUS	INC HL	; point HL register pair to 0C81H
0CA0 70		LD (HL),B	; store B in memory location ; 0C81H
0CA1 76		HALT	; program end

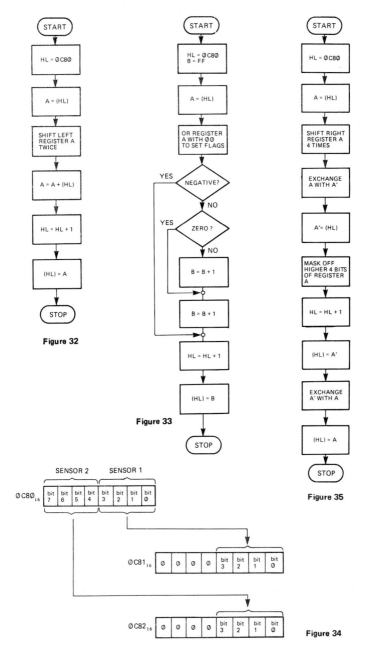

Figure 32

Figure 33

Figure 35

Figure 34

127

Algorithm:

(i)   load register A (accumulator) with the contents of memory location $\emptyset C8\emptyset_{16}$;

(ii) shift the contents of register A four places to the right and store the result in memory location $\emptyset C82_{16}$;

(iii) load register A again with data from memory location $\emptyset C8\emptyset_{16}$;

(iv) logical AND register A with $\emptyset F_{16}$ to mask off the most significant four bits, and store the result in memory location $\emptyset C81_{16}$; and

(v)   stop.

Flowchart: see *Figure 35*.

Program code:

Machine code	DATA SENS1 SENS2	EQU $\emptyset C8\emptyset$H EQU $\emptyset C81$H EQU $\emptyset C82$H ORG $\emptyset C9\emptyset$H	Assembly language
$\emptyset C9\emptyset$ 21 8$\emptyset$ $\emptyset$C		LD HL,DATA	; point HL register pair to $\emptyset C8\emptyset$H
$\emptyset C93$ 7E		LD A,(HL)	; load register A with data word
$\emptyset C94$ CB 3F		SRL A	
$\emptyset C96$ CB 3F		SRL A	
$\emptyset C98$ CB 3F		SRL A	; shift higher 4 bits into lower
$\emptyset C9A$ CB 3F		SRL A	; 4 bit positions
$\emptyset C9C$ $\emptyset 8$		EX AF,AF'	; swap registers
$\emptyset C9D$ 7E		LD A,(HL)	; load register A with data word
$\emptyset C9E$ E6 $\emptyset$F		AND $\emptyset$FH	; mask off higher 4 bits
$\emptyset CA\emptyset$ 23		INC HL	; increment HL register pair by 1
$\emptyset CA1$ 77		LD (HL),A	; store sensor 1 data in $\emptyset C81$H
$\emptyset CA2$ 23		INC HL	; increment HL register pair by 1
$\emptyset CA3$ $\emptyset 8$		EX AF,AF'	; swap registers
$\emptyset CA4$ 77		LD (HL),A	; store sensor 2 data in $\emptyset C82$H
$\emptyset CA5$ 76		HALT	; program end

Algorithm:

(i)   load accumulator A with the contents of memory location $\emptyset\emptyset 8\emptyset_{16}$;

(ii) load accumulator B with the contents of memory location $\emptyset\emptyset 81_{16}$;

(iii) store the contents of accumulator A in memory location $\emptyset\emptyset 81_{16}$;

(iv) store the contents of accumulator B in memory location $\emptyset\emptyset 8\emptyset_{16}$;

(v)   stop.

Flowchart: see *Figure 36*.

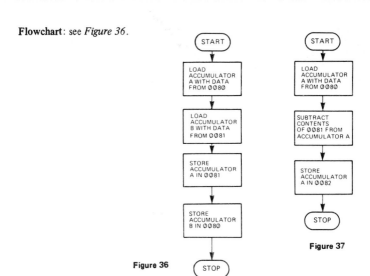

Figure 36

Figure 37

Program code:

Machine code	DATA1	EQU $0080	Assembly language
	DATA2	EQU $0081	
		ORG $0020	
0020 96 80		LDA A DATA1	; load accumulator A from $0080
0022 D6 81		LDA B DATA2	; load accumulator B from $0081
0024 97 81		STA A DATA2	; store accumulator A in $0081
0026 D7 80		STA B DATA1	; store accumulator B in $0080
0028 3F		SWI	; program end

Problem 26 Construct a trace table for the program written in *Problem 25*.

For the purpose of constructing this trace table it is assumed that memory
locations 0080_{16} and 0081_{16} contain 03_{16} and 05_{16} initially.

Address bus	Instruction mnemonic	Acc		Memory locations		Flag register				Program counter
		A	B	0080	0081	C	N	V	Z	
		XX	XX	03	05	X	X	X	X	0020
0020	LDAA $80	03	XX	03	05	X	0	0	0	0020
0022	LDAB $81	03	05	03	05	X	0	0	0	0024
0024	STAA $81	03	05	03	03	X	0	0	0	0026
0026	STAB $80	03	05	05	03	X	0	0	0	0028
0028	SWI									

Algorithm:
(i)   load accumulator A with data from memory location 0080_{16};
(ii) subtract the data in memory location 0081_{16} from accumulator A;
(iii) store the difference in memory location 0082_{16};
(iv) stop.

Flowchart: see *Figure 37*.

Program code:

Machine code	DATA1	EQU $0080	Assembly language
	DATA2	EQU $0081	
	DIFF	EQU $0082	
		ORG $0020	
0020 96 80		LDA A DATA1	; get minuend
0022 90 81		SUB A DATA2	; subtract subtrahend
0024 97 82		STA A DIFF	; store difference in location
			; $0082
0026 3F		SWI	; program end

For the purpose of constructing this trace table it is assumed that memory locations 0080_{16} and 0081_{16} contain 09_{16} and 06_{16} initially.

Address bus	Instruction mnemonic	Acc A	Memory locations			Flag register				Program counter
			0080	0081	0082	C	N	V	Z	
		XX	09	06	XX	X	X	X	X	0020
0020	LDAA $80	09	09	06	XX	X	0	0	0	0022
0022	SUBA $81	03	09	06	XX	0	0	0	0	0024
0024	STAA $82	03	09	06	03	0	0	0	0	0026
0026	SWI									

Algorithm:

(i) load accumulator A with the hex. number to be converted;

(ii) compare the value in accumulator A with $0A_{16}$;

(iii) if the N-flag is set (N = 1), the number is in the range 00_{16} to 09_{16} and no adjustment is necessary;

(iv) if the N-flag is clear (N = 0), the number is in the range $0A_{16}$ to $0F_{16}$ and 06_{16} must be added;

(v) store the decimal equivalent of the number in memory location 0081_{16};

(vi) stop.

Flowchart: see *Figure 38*.

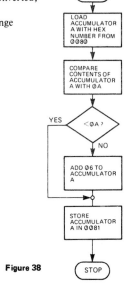

Figure 38

Program code:

Machine code		Assembly language	
	HEXNUM	EQU $0080	
	DECNUM	EQU $0081	
		ORG $0020	
0020 96 80		LDA A HEXNUM	; load accumulator A with hex ; number
0022 81 0A		CMP #$0A	; compare with $0A
0024 2B 02		BMI SAVE	; do not adjust number
0026 8B 06		ADD A #$06	; convert to decimal
0028 97 81	SAVE	STA DECNUM	; store decimal equivalent in ; $0081
002A 3F		SWI	; program end

Problem 30 Memory location 0080_{16} contains a hexadecimal number in the range 00_{16} to 50_{16}. Write a program (including algorithm and flowchart), using 6800 code, which multiplies this number by 5 and stores the product in memory location 0081_{16}. The program may be located in any convenient area of memory.

Algorithm:

(i) load accumulator A with the contents of memory location 0080_{16};

(ii) shift the contents of accumulator A two places to the left (equivalent to multiplying by 4);

(iii) add the contents of memory location 0080_{16} to the accumulator (with (ii), is equivalent to multiplying by 5);

(iv) store the contents of accumulator A in memory location 0081_{16}, and

(v) stop.

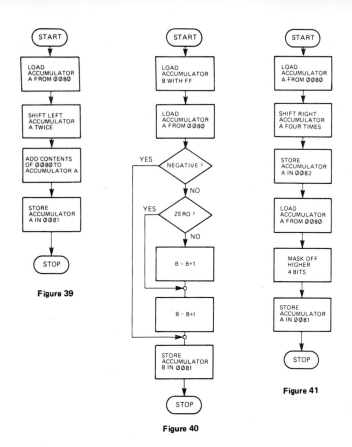

Figure 39

Figure 40

Figure 41

Flowchart: see *Figure 39*.

Program code:

Machine code				Assembly language
	HEXNUM	EQU	$0080	
	PROD	EQU	$0081	
		ORG	$0020	
0020 96 80			LDA A HEXNUM	; load accumulator A with hex.
				; number
0022 48			ASL A	
0023 48			ASL A	; shift left accumulator A twice
0024 9B 80			ADD A HEXNUM	; add hex. number to accumula-
				; tor A
0026 97 81			STA A PROD	; store the product in location
				; $0081
0028 3F			SWI	; program end

132

Program 31   Write a program (including algorithm and flowchart), using 6800 code, which checks a data word in memory location 0080_{16} and stores a value (flag) in memory location 0081_{16} with the following meaning:
(i) FF_{16} if the data word is negative,
(ii) 00_{16} if the data word is zero, or
(iii) 01_{16} if the data word is positive.
The program may be located in any convenient area of memory.

Algorithm:
(i)   load accumulator B with FF_{16};
(ii) load accumulator A with data from memory location 0080_{16};
(iii) if the data word is negative, store the contents of accumulator B in memory location 0081_{16} and stop;
(iv) if the data word is zero, increment accumulator B by 1 and store its contents in memory location 0081_{16} and stop;
(v)   if the data word is positive, increment accumulator B by 2 and store its contents in memory location 0081_{16} and stop.

Flowchart: see *Figure 40*.

Program code:

Machine code		WORD	EQU $0080	Assembly language
		FLAG	EQU $0081	
			ORG $0020	
0020 C6 FF			LDA B #$FF	; load accumulator B with $FF
0022 96 80			LDA A WORD	; load accumulator A with data ; word
0024 2B 04			BMI MINUS	; test for negative data word
0026 27 01			BEQ ZERO	; test for zero data word
0028 5C			INC B	; must be positive by default
0029 5C		ZERO	INC B	; increment B as appropriate
002A D7 81		MINUS	STA B FLAG	; store B in memory location ; $0081
002C 3F			SWI	; program end

Problem 32   Two electrical sensors in a control system provide two 4-bit words which are stored together in a single memory location (at 0080_{16}) as a single 8-bit word, as illustrated in *Figure 42*. Write a program (including algorithm and flowchart), using 6800 code, to separate out (unpack) the two 4-bit words and store them in two separate memory locations, 0081_{16} and 0082_{16}, such that they occupy the lower four bits in each of their respective memory locations.

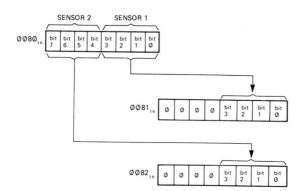

Figure 42

Algorithm:

(i) load accumulator A with the contents of memory location 0080_{16};

(ii) shift accumulator A contents right 4 times and store the result in memory location 0082_{16};

(iii) load accumulator A again with data from memory location 0080_{16};

(iv) logical AND accumulator A with $0F_{16}$ to mask off the most significant four bits, and store the result in memory location 0081_{16};

(v) stop.

Flowchart: see *Figure 41*.

Program code:

Machine code		DATA	EQU $0080	Assembly language
		SENS1	EQU $0081	
		SENS2	EQU $0082	
			ORG $0020	
0020	96 80		LDA A DATA	; load accumulator A with data
0022	44		LSR A	
0023	44		LSR A	
0024	44		LSR A	; shift higher 4 bits into lower
0025	44		LSR A	; 4 bit positions
0026	97 82		STA A SENS2	; store sensor 2 data in $0082
0028	96 80		LDA A DATA	; reload accumulator A with data
002A	84 0F		AND A #$0F	; mask off higher 4 bits
002C	97 81		STA A SENS 1	; store sensor 1 data in $0081
002E	3F		SWI	; program end

C FURTHER PROBLEMS ON MICROPROCESSOR INSTRUCTION SETS AND MACHINE CODE PROGRAMS

(a) SHORT ANSWER PROBLEMS

1 The symbolic notation $A + 1 \rightarrow A$ represents an . . . . instruction.

2 The symbolic notation $A + M + C \rightarrow A,C$ represents an . . . . instruction.

3 An instruction whose function is to cause a microprocessor program counter to be loaded with a new value is called a . . . . instruction.

4 An instruction which multiplies the contents of a microprocessor accumulator by two is called an . . . . instruction.

5 An instruction which divides the contents of a microprocessor accumulator by two is called a . . . . instruction.

6 The only effect of a 'no operation' (NOP) instruction is to . . . .
. . . .

7 NOP, SEC, TAX and TXA are all examples of instructions which use an . . . .
mode of addressing.

8 Microprocessor instruction sets may be divided into three groups known as
. . . . , ... and . . . . groups.

9 One advantage of using the relative addressing mode is that it makes programs readily . . . .

10 A microprocessor instruction which subtracts the contents of a specified memory location from the accumulator, but does not store the difference back in the accumulator is called a . . . . instruction.

11 A microprocessor instruction which performs a logical AND between a specified memory location and the accumulator, but does not store the result of this operation back into the accumulator is called a . . . . instruction.

12 The maximum range of a branch when using relative addressing is between . . . .
and . . . . relative to the current value of the program counter.

13 If the second byte of a 'load accumulator' instruction contains the actual data to be loaded into the accumulator, the . . . . mode of addressing is being used.

14 An instruction to branch to location 0130_{16} is located in memory at 0120_{16}. The correct offset to store in location 0121_{16} is . . . .

15 Memory locations in the range 0000_{16} to $00FF_{16}$ are frequently referred to as
. . . . addresses.

Reference to the appropriate instruction set in Appendix B may be necessary to answer problems 16 to 24.

16 In order to load the accumulator of a 6502 microprocessor with data stored in memory location 0080_{16}, the correct opcode to use is . . . .

17 In order to store the contents of the accumulator of a 6502 microprocessor in memory location 0200_{16}, the correct opcode to use is . . . .

18 In order to load index register X of a 6502 microprocessor with a constant, the correct opcode to use is . . . .

19 In order to load accumulator A of a 6800 microprocessor with data stored in memory location 0080_{16}, the correct opcode to use is . . . .

20 In order to store the contents of accumulator A of a 6800 microprocessor in memory location 0200_{16}, the correct opcode to use is . . . .

21 In order to load index register X of a 6800 microprocessor with a constant, the correct opcode to use is . . . .

22 In order to load register A of a Z80 microprocessor with a constant, the correct opcode to use is . . . .

23 In order to load register A of a Z80 microprocessor with the contents of the memory location pointed to by the HL register pair, the correct opcode to use is . . . .

24 In order to transfer the contents of register B to register A of a Z80 microprocessor, the correct opcode to use is . . . .

25 A sequence of steps which define the solution to a problem is called an . . . .

26 A graphical method for representing the solution to a problem is called . . . .

27 A computer program which converts instructions in mnemonic form into machine executable instructions is called an . . . .

28 Microprocessor instructions which are written in binary code are called . . . .

29 The removal of errors from a microprocessor program is a process which is called . . . .

30 A list which shows the contents of each register of a microprocessor after the execution of each instruction is called a . . . .

31 A program in firmware, resident in a microcomputer system, which enables programs to be entered via a keyboard is called a . . . . program.

32 The first step in writing a microcomputer program is to write a . . . .

33 In assembly language programs, names given to memory locations are known as . . . .

34 The lack of an algorithm indicates that a problem has . . . .

(b) CONVENTIONAL PROBLEMS

1 Rewrite worked problem 1 by making use of index registers X and Y so that a program of reduced length results.

2 Construct a trace table for *Worked Problem 13*.

3 Construct a trace table for *Worked Problem 14*.

4 Two 16-bit numbers are located in four consecutive memory locations starting at 0080_{16}. Write a program (including algorithm and flowchart), using 6502 code, to add together these two numbers. The sum may be stored in any convenient memory location. State why it may be necessary to reserve **three** memory locations for the result.

5 Construct a trace table for *Worked Problem 15*.

6 Construct a trace table for *Worked Problem 16*.

7 Memory locations $\emptyset\emptyset 8\emptyset_{16}$ and $\emptyset\emptyset 81_{16}$ contain two 4-bit BCD numbers in each of their lower 4 bits. Write a program (including algorithm and flowchart), using 6502 code, to pack these two BCD numbers into one memory location (any convenient location may be used).

8 Three numbers are located in memory at $\emptyset\emptyset 8\emptyset_{16}$, $\emptyset\emptyset 81_{16}$ and $\emptyset\emptyset 82_{16}$. Write a program (including algorithm and flowchart), using 6502 code, to arrange these numbers in ascending order (smallest number in $\emptyset\emptyset 8\emptyset_{16}$).

9 Construct a trace table for *Worked Problem 21*.

10 Construct a trace table for *Worked Problem 22*.

11 Two 16-bit numbers are located in four consecutive memory locations starting at $\emptyset C8\emptyset_{16}$. Write a program (including algorithm and flowchart), using Z80 code, to add together these two numbers. The sum may be stored in any convenient memory location. State why it is necessary to reserve **three** memory locations for the result.

12 Construct a trace table for *Worked Problem 23*.

13 Construct a trace table for *Worked Problem 24*.

14 Memory locations $\emptyset C8\emptyset_{16}$ and $\emptyset C81_{16}$ contain two 4-bit BCD numbers in each of their lower 4 bits. Write a program (including algorithm and flowchart), using Z80 code, to pack these two BCD numbers into one memory location (any convenient location may be used).

15 Three numbers are located in memory at $\emptyset C8\emptyset_{16}$, $\emptyset C81_{16}$ and $\emptyset C82_{16}$. Write a program (including algorithm and flowchart), using Z80 code, to arrange these numbers in ascending order (smallest number in $\emptyset C8\emptyset_{16}$).

16 Construct a trace table for *Worked Problem 29*.

17 Construct a trace table for *Worked Problem 30*.

18 Two 16-bit numbers are located in four consecutive memory locations starting at $\emptyset\emptyset 8\emptyset_{16}$. Write a program (including algorithm and flowchart), using 6800 code, to a add together these two numbers. The sum may be stored in any convenient memory location. State why it may be necessary to reserve **three** memory locations for the result.

19 Construct a trace table for *Worked Problem 31*.

20 Construct a trace table for *Worked Problem 32*.

21 Memory locations $\emptyset\emptyset 8\emptyset_{16}$ and $\emptyset\emptyset 81_{16}$ contain two 4-bit BCD numbers in each of their lower 4 bits. Write a program (including algorithm and flowchart), using 6800 code, to pack these two BCD numbers into one memory location (any convenient location may be used).

22 Three numbers are located in memory at $\emptyset\emptyset 8\emptyset_{16}$, $\emptyset\emptyset 81_{16}$ and $\emptyset\emptyset 82_{16}$. Write a program (including algorithm and flowchart), using 6800 code, to arrange these numbers in ascending order (smallest number in $\emptyset\emptyset 8\emptyset_{16}$).

5 Programs with loops

A MAIN POINTS CONCERNED WITH PROGRAMS WITH LOOPS

1   There is often a requirement in a system to perform a given task several times in succession. For example, a system may be required to generate a sequence of four identical pulses in quick succession. It is possible to repeat that section of a program dealing with a particular task as many times as necessary, but this generally results in longer programs than necessary, and uneconomical use of available memory. Depending upon the complexity of a task, a better solution may be to repeatedly pass through the same section of a program the required number of times. This technique is known as 'looping'. A loop may be defined as the body of a program which is repeated a given number of times. A comparison between these two techniques, using the generation of four pulses as an example, is illustrated in *Figure 1(a) and (b)*.

2   Certain types of program are based on a **continuous loop**, such that when the end of a program is reached, a jump back to the start occurs and provides continuous operation until such time that the microcomputer is reset or switched off. Other programs require a **fixed number of passes** around a loop to occur. In such programs a register or memory location is used as a **loop counter** which is incremented or decremented by one after each pass through the loop. The value in a loop counter is tested after each pass through the loop to determine whether further looping is to occur. A program of this type has three sections which are:

   (i)   an **initialisation section** (set up loop counter);
   (ii) a **processing section** (program body);
   (iii) a **loop control section**.

These sections are organised as illustrated in *Figure 2(a) and (b)*.

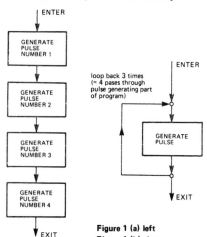

Figure 1 (a) left
Figure 1 (b) above

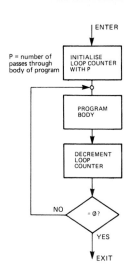

Figure 2 (a) above
Figure 2 (b) right

3   In many programs, although the same process is repeated many times, the data processed is different on each pass through the loop. In these cases, the operand of an instruction within the loop must be changed after each pass through the loop. For example, consider a program which is used to store data $\emptyset\emptyset_{16}$ in every memory location from address $\emptyset\emptyset\emptyset\emptyset_{16}$ to $\emptyset\emptyset\emptyset F_{16}$ inclusive. In this example, the operand of a 'store accumulator in memory' instruction must be changed after each pass through the loop. For this purpose, **indexed addressing** may be used as shown in *Figure 3(a) and (b)* for the 6502 and 6800 microprocessors, or **register indirect addressing** as shown in *Figure 3(c)* for the Z80 microprocessor.

4   A **jump** or **branch instruction** is used to close a program loop, and its operand must be of such a value as to enable the program counter to locate the next instruction in the loop. Two addressing modes are commonly used for these instructions:

    (i)   **absolute** or **extended**, in which the actual address of the next instruction to be executed is used as the operand; or

    (ii)   **relative**, in which the operand is a two's complement offset which is added to the current value of the program counter to form the address of the next instruction to be executed (see *Worked Problem 2*, Chapter 4).

5   As an example of the use of looping techniques, consider the design of a program which adds together five numbers stored in five consecutive memory locations and stores the sum in a sixth location:

Algorithm:
    (i)   zero the accumulator (which keeps a running total of the sum);
    (ii)   load an index register with the number of values to be totalled;
    (iii)   select a value from the table of five numbers to be added, using indexed addressing, and add to the accumulator;

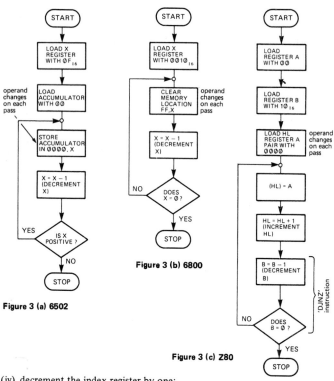

Figure 3 (a) 6502

Figure 3 (b) 6800

Figure 3 (c) Z80

(iv) decrement the index register by one;

(v) test the value in the index register to determine if it is decremented to zero, and if not, repeat steps (iii) to (v);

(vi) store the total (which is in the accumulator) in memory,

(vii) stop.

Flowchart: see *Figure 4*

Program code:

(a) 6502

Machine code		Assembly language
	TABLE=$0080	
	SUM=$0086	
	*=$0020	
0020 A9 00	LDA #$00	; clear accumulator
0022 A2 05	LDX #$05	; X=number of values totalled
0024 18	CLC	
0025 75 80	TOTAL ADC TABLE,X	; select a value from the table
0027 CA	DEX	; decrement register X by 1
0029 D0 FB	BNE TOTAL	; loop for next addition
002B 85 86	STA SUM	; save total in location $0086
002D 00	BRK	; program end

In this example, the five numbers to be totalled are assumed to be stored in memory locations 0081_{16} to 0085_{16} inclusive. Although the table of numbers starts at location 0080_{16}, the value in this location is not added to the total since the program breaks out of the loop when index register X has a value of zero.

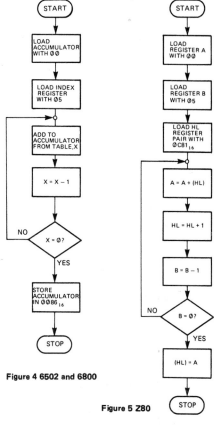

Figure 4 6502 and 6800

Figure 5 Z80

(b) 6800

Machine code		Assembly language
	TABLE EQU $0080	
	SUM EQU $0086	
	ORG $0020	
0020 4F	CLR A	; clear accumulator
0021 CE 00 05	LDX $0005	; X=number of values totalled
0024 AB 80	TOTAL ADD A TABLE,X	; select a value from the table
0026 09	DEX	; decrement register X by 1
0027 26 FB	BNE TOTAL	; loop for next addition
0029 97 86	STA A SUM	; save total in location $0086
002B 3F	SWI	; program end

In this example, the five numbers to be totalled are assumed to be stored in memory locations 0081_{16} to 0085_{16} inclusive. Although the table of numbers starts at location 0080_{16}, the value in this location is not added to the total since the program breaks out of the loop when index register X has a value of zero.

141

(c) Z80

A Z80 microprocessor has many different ways of achieving the same result, due to its very comprehensive instruction set. Indexed addressing could be used for adding the five numbers (in a similar manner to that already used with the 6502 and 6800 microprocessors), but this form of addressing results in a longer program than is necessary. A main disadvantage of a Z80 microprocessor is that many operations do **not** control the status register flags. In particular, incrementing or decrementing Z80 index registers has no effect on any flags, so it becomes more difficult to determine when to break out of a program loop when using indexed addressing.

For this problem, a different approach is adopted. The HL register pair is used as a memory pointer which is incremented after each addition, and register B is used as a loop counter. A single instruction, 'DJNZ' (decrement and jump if not zero), is used to decrement register B and loop back for a further addition if register B does not contain a value of zero. The flowchart for this method is shown in *Figure 5*.

Machine code		TABLE	EQU 0C81H	Assembly language
			ORG 0C90H	
0C90 AF			XOR A	; clear register A (accumulator)
0C91 06 05			LD B,05H	; set up loop counter
0C93 21 81 0C			LD HL, TABLE	; point HL to first number in ; table
0C96 86		TOTAL	ADD A,(HL)	; add a value from the table
0C97 23			INC HL	; point HL to next number
0C98 10 FC			DJNZ TOTAL	; loop for next addition
0C9A 77			LD (HL),A	; save total in 0C86H
0C9B 76			HALT	; program end

In this example, the five numbers to be totalled are assumed to be stored in memory locations $0C81_{16}$ to $0C85_{16}$ inclusive. Note the use of EX-ORing the accumulator with itself as a means of clearing the accumulator.

B WORKED PROBLEMS

Problem 1   A block of 32_{10} numbers is stored in memory starting at address 0080_{16}. Write a program (including algorithm and flowchart), using 6502 code, to move the block of numbers to a new starting address of $00B0_{16}$.

Algorithm:
(i) set up index register X as a loop counter by loading it with $1F_{16}$ (31_{10});
(ii) load the accumulator with a number from the block, using indexed addressing (index register X);

(iii) store the contents of the accumulator in the new location, using indexed addressing (index register X);

(iv) decrement index register X by 1, and repeat (ii) and (iii) if its value is positive (zero counts as positive);

(v) stop.

Flowchart: see *Figure 6*.

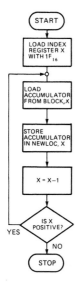

Figure 6

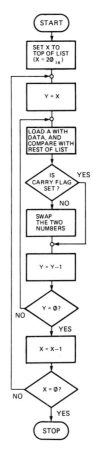

Figure 7

143

Program code:

Machine code	BLOCK NEWLOC *	=$0080 =$00B0 =$0020	Assembly language	
0020 A2 1F			LDX #$1F	; set up loop counter
0022 B5 80	LOOP		LDA BLOCK,X	; select a number from the block
0024 95 C0			STA NEWLOC,X	; move it to new location
0026 CA			DEX	; decrement X by 1
0027 10 F9			BPL LOOP	; repeat until whole block is ; moved
0029 00			BRK	; program end

Note that this program copies the block of numbers into new locations and does not change the original block of numbers in any way.

Problem 2 A list of 32_{10} numbers is stored in memory, starting at address 0080_{16}. Write a program (including algorithm and flowchart), using 6502 code, to **sort** these numbers into ascending numerical order (all numbers are regarded as unsigned, i.e. all positive, for the purpose of this program).

Algorithm: One possible method for sorting numbers is known as a '**bubble sort**', and it operates as follows:

(i) select the last number in the list and compare its magnitude with all other numbers in the list, starting with the next to last number, and finishing with the first number;

(ii) swap the numbers after each comparison, if necessary, so that after all numbers are compared, the largest number occupies the last position in the list (i.e. the largest 'bubbles' to the top);

(iii) select the next to last number from the list and repeat (i) and (ii) for all numbers lower down the list;

(iv) repeat (iii) for all numbers in the list, by which time the numbers are sorted into ascending numerical order;

(v) stop.

Flowchart: see *Figure 7*.

Program code:

Machine code			Assembly language
	TEMP	=$007F	
	LIST	=$0080	
	.*	=$0020	
0020 A2 20		LDX #$20	; load X with number of values
0022 8A	LAST	TXA	
0023 A8		TAY	; make Y=X
0024 B5 80	NEXT	LDA LIST,X	; get number at end of list
0026 D9 7F 00		CMP LIST-1,Y	; compare with others in list
0029 B0 0C		BCS NOSWAP	; numbers in correct order
002B 85 7F		STA TEMP	; numbers must be swapped
002D B9 7F 00		LDA LIST-1,Y	; start of swap routine
0030 95 80		STA LIST,X	
0032 A5 7F		LDA TEMP	
0034 99 7F 00		STA LIST-1,Y	; swap completed
0037 88	NOSWAP	DEY	; point Y further down list
0038 D0 EA		BNE NEXT	; repeat until bottom of list
003A CA		DEX	; point X further down list
003B D0 E5		BNE LAST	; repeat until list is sorted
003D 00		BRK	; program end

Problem 3 Two 8-bit numbers are stored in memory locations 0080_{16} and 0081_{16}. Write a program (including algorithm and flowchart), using 6502 code, to **multiply** these two numbers together and store the product in memory locations 0082_{16} and 0083_{16}.

Algorithm: There are many different ways of multiplying numbers. One convenient method is to 'shift and add' in a similar manner to that used for multiplying decimal numbers, and this method may be summarised as follows:
(i) clear the memory location reserved for the product;
(ii) shift the multiplier to the right and test the state of the carry flag;
(iii) if the carry flag is set, add the multiplicand to the contents of the memory location reserved for the product (the contents of this location are known as the 'partial product' until multiplication is completed);
(iv) if the carry flag is clear, add nothing to the partial product;
(v) shift the multiplicand to the left one place;
(vi) repeat (ii) to (v) a total of eight times;
(vii) stop.

Note, in the expression $X \times Y = Z$, X is called the **multiplicand**, Y is called the **multiplier** and Z is called the **product**.

This process is illustrated in *Figure 8*. Note, the multiplicand is 8-bits wide and is shifted to the left a total of eight times. Therefore it is necessary to reserve a memory location to retain the bits of the multiplicand as they are shifted out of their initial location. Also, an 8-bit by 8-bit multiplication may give a product of up to 16 bits, therefore two memory locations are reserved for storing the product.

Flowchart: see *Figure 9*.

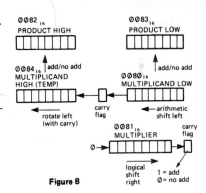

Figure 8

Program code:

Machine code	MCAND =$0080 MPLIER =$0081 PRODHI =$0082 PRODLO=$0083 TEMP=$0084 *=$0020		Assembly language
0020 A9 00		LDA #$00	; clear accumulator
0022 85 82		STA PRODHI	; clear product high byte location
0024 85 83		STA PRODLO	; clear product low byte location
0026 85 84		STA TEMP	; clear temporary storage location
0028 A2 08		LDX #$08	; X=loop counter (8 shift & add)
002A 46 81	SHIFT	LSR MPLIER	; shift multiplier right one place
002C 90 0D		BCC NOADD	; add nothing if carry clear
002E 18		CLC	; prepare for addition
002F A5 83		LDA PRODLO	; get partial product low byte
0031 65 80		ADC MCAND	; add on multiplicand low byte
0033 85 83		STA PRODLO	; and put it back
0035 A5 82		LDA PRODHI	; get partial product high byte
0037 65 84		ADC TEMP	; add on multiplicand high byte
0039 85 82		STA PRODHI	; and put it back
003B 06 80	NOADD	ASL MCAND	; shift multiplicand left one place
003D 26 84		ROL TEMP	; rotate carry flag into TEMP
003F CA		DEX	; decrement X by 1
0040 D0 E8		BNE SHIFT	; shifted 8 times?, if not loop back
0042 00		BRK	; program end

Problem 4 A block of 32_{10} numbers is located in memory, starting at address 0080_{16}. Write a program (including algorithm and flowchart), using 6502 code, to find the **largest number** in the block and store this value in memory location $00A0_{16}$.

Algorithm:
(i)   load index register X with the length of the block,
(ii)   load index register Y with the last number in the block,
(iii) compare the remaining numbers in the block with the value in index register Y,
(iv)   if a number larger than that which is in index register Y is found, reload Y with this number,
(v)   decrement index register X by 1 after each comparison so that it is known when the first number in the block is reached and no further comparisons are necessary,
(vi)   store index register Y contents in memory at location $00A0_{16}$,
(vii) stop.

Flowchart: see *Figure 10*.

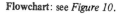

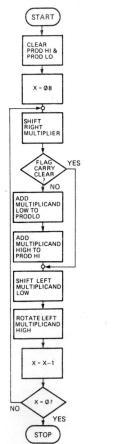

Figure 9

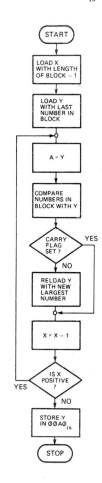

Figure 10

147

Program code:

Machine code	BLOCK MAX *	=$0080 =$00A0 =$0020	Assembly language
0020 A2 1F		LDX #$1F	; load X with number of values
0022 A4 9F		LDY BLOCK+$1F	; load Y with last number in block
0024 98	COMP	TYA	; transfer Y to A for comparison
0025 D5 80		CMP BLOCK,X	; compare A with rest of block
0027 B0 02		BCS NEXT	; if A holds larger value, go to ; NEXT
0029 B4 80		LDY BLOCK,X	; load Y with new larger value
002B CA	NEXT	DEX	; point to next number in block
002C 10 F6		BPL COMP	; repeat if not at end of block
002E 84 A0		STY MAX	; store maximum number in $00A0
0030 00		BRK	; program end

Problem 5 A list of non-zero numbers is stored in memory starting at location 0080_{16}, and is terminated with 00_{16}. Write a program (including algorithm and flowchart), using 6502 code, to determine the **length** of the list and store this in memory location $007F_{16}$.

Algorithm:

(i)   zero index register X;

(ii) load the accumulator with the first number in the list, using indexed addressing (index register X);

(iii) check the value in the accumulator to see if it is zero;

(iv) if the accumulator contains a non-zero value, increment index register X by 1 and repeat (ii) to (iv);

(v)   if the accumulator contains zero, store index register X in memory location $007F_{16}$;

(vi) stop.

Flowchart: see *Figure 11*.

Program code:

Machine code	LENGTH LIST *	=$007F =$0080 =$0020	Assembly language
0020 A2 00		LDX #$00	; clear index register X
0022 B5 80	CHECK	LDA LIST,X	; get a number from the list
0024 F0 03		BEQ SAVE	; if zero, end of list reached
0026 E8		INX	; increment number counter ; (X) by 1
0027 D0 F9		BNE CHECK	; not at end of list yet
0029 86 7F	SAVE	STX LENGTH	; save list length in $007F
002B 00		BRK	; program end

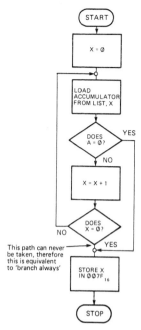

START

X = Ø

LOAD
ACCUMULATOR
FROM LIST, X

DOES
A = Ø?   YES

NO

X = X + 1

DOES
X = Ø?

NO   YES

This path can never
be taken, therefore
this is equivalent
to 'branch always'

STORE X
IN ØØ7F₁₆

STOP

Figure 11

Problem 6 Two 8-bit numbers are stored in memory locations ØØ8Ø_{16} and ØØ81_{16}. Write a program (including algorithm and flowchart), using 6502 code, to **divide** the contents of ØØ8Ø_{16} by the contents of ØØ81_{16} and store the quotient in location ØØ82_{16} and any remainder in location ØØ83_{16}.

Algorithm:
(i)   clear the accumulator;
(ii) shift the dividend one place to the left and shift any carry into the accumulator;
(iii) subtract the divisor from the accumulator;
(iv) if the carry flag is clear (divisor larger than accumulator), restore the accumulator by adding on the divisor;
(v)   rotate the quotient to the left, thus shifting in the current state of the carry flag;
(vi) carry out (ii) to (v) a total of eight times;
(vii) the value left in the accumulator is the remainder which is stored in location ØØ83_{16};
(viii) stop.

This process is illustrated in *Figure 12(a)*

Note, in the expression $X \div Y = Z$, X is called the
dividend, Y is called the **divisor** and Z is called the **quotient**.

Flowchart: see *Figure 12(b)*.

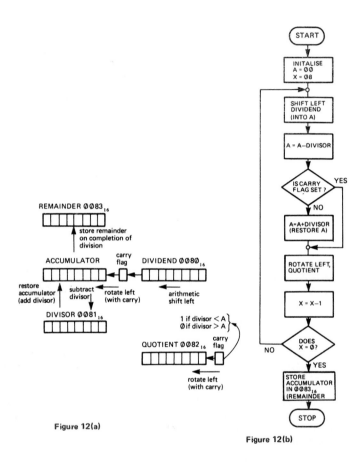

REMAINDER 0083₁₆

store remainder on completion of division

ACCUMULATOR   carry flag   DIVIDEND 0080₁₆

restore accumulator (add divisor)   subtract divisor   rotate left (with carry)   arithmetic shift left

DIVISOR 0081₁₆

1 if divisor < A
0 if divisor > A

carry flag

QUOTIENT 0082₁₆

rotate left (with carry)

Figure 12(a)

START

INITALISE
A = 00
X = 08

SHIFT LEFT
DIVIDEND
(INTO A)

A = A−DIVISOR

IS CARRY
FLAG SET ? —— YES

NO

A=A+DIVISOR
(RESTORE A)

ROTATE LEFT,
QUOTIENT

X = X−1

DOES
X = 0? —— NO

YES

STORE
ACCUMULATOR
IN 0083₁₆
(REMAINDER)

STOP

Figure 12(b)

Program code:

Machine code		Assembly language	
	DIVDND	=$0080	
	DIVSOR	=$0081	
	QUOTNT	=$0082	
	REMNDR	=$0083	
	*	=$0020	
0020 A9 00		LDA #$00	; clear accumulator
0022 A2 08		LDX #$08	; set up X as a loop counter
0024 06 80	SHIFT	ASL DIVDND	; shift left DIVDND and
			; accumulator
0026 2A		ROL A	; as a 16-bit register
0027 38		SEC	; prepare for subtraction
0028 E5 81		SBC DIVSOR	; subtract DIVSOR from
			; accumulator
002A B0 03		BCS SET	; is DIVSOR greater than
			; accumulator?
002C 65 81		ADC DIVSOR	; if so, restore accumulator
002E 18		CLC	; clear carry flag,
002F 26 82	SET	ROL QUOTNT	; and shift a digit into the quotient
0031 CA		DEX	; decrement X by 1
0032 D0 F0		BNE SHIFT	; repeat if not done 8 times
0034 85 83		STA REMNDR	; save the remainder
0036 00		BRK	; program end

Problem 7 A block of 32_{10} numbers is stored in memory starting at address $0C90_{16}$. Write a program (including algorithm and flowchart), using Z80 code, to **move** the block of numbers to a new starting address of $0CC0_{16}$.

Algorithm:

This routine is known as a '**block transfer**', and, unlike most other microprocessors, a Z80 has a special instruction for this type of operation – **LDIR (load, increment and repeat)**. This instruction moves data from a location pointed to by the HL register pair to a new location pointed to by the DE register pair. HL and DE register pairs are both incremented by one, and the BC register pair is decremented by one and further moves take place until BC register pair is reduced to zero. Therefore, to accomplish a block move with a Z80 microprocessor, the following steps are necessary:

(i) load HL register pair with the starting address of the block to be moved;
(ii) load DE register pair with the address of the start of the new location;
(iii) load BC register pair with the number of bytes to be moved;
(iv) issue an 'LDIR' instruction;
(v) stop.

Flowchart: see *Figure 13*.

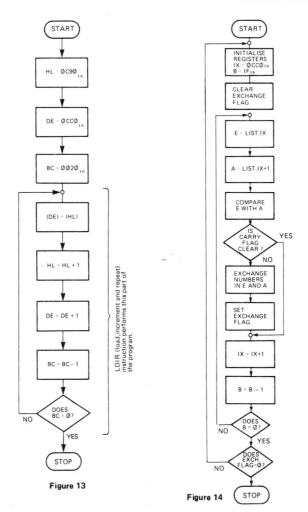

Figure 13

Figure 14

Program code:

Machine code	BLOCK	EQU 0C90H	Assembly language
	NEWLOC	EQU 0CC0H	
		ORG 0C80H	
0C80 21 90 0C		LD HL,BLOCK	; point HL pair to start of block
0C83 11 C0 0C		LD DE,NEWLOC	; point DE pair to start of new block
0C86 01 20 00		LD BC,20H	; put number of bytes to move in BC
0C89 ED B0		LDIR	; transfer the block
0C8B 76		HALT	; program end

152

Problem 8 A list of 32_{10} numbers is stored in memory, starting at address $0CC0_{16}$. Write a program (including algorithm and flowchart), using Z80 code, to **sort** these numbers into ascending numerical order (all numbers are regarded as unsigned, i.e. all positive, for the purpose of this program).

Algorithm: A version of the '**bubble sort**' described in *Worked Problem 2* may be used, and operates as follows:

(i)   select the first number from the list and compare its magnitude with the second number in the list;

(ii)   if the second number is larger than the first number, leave unchanged;

(iii)   if the second number is smaller than the first number, exchange the positions of the two numbers in the list and set an 'exchange flag' to indicate that an exchange takes place;

(iv)   repeat (i) to (iii) with the second and third numbers in the list, third and fourth etc., until the end of the list is reached;

(v)   repeat (i) to (iv) until no further exchanges take place, determined by inspection of the '**exchange flag**';

(vi)   stop.

Flowchart: see *Figure 14*.

Program code:

Machine code		LIST	EQU 0CC0H	Assembly language
			ORG 0C90H	
0C90	06 1F	SORT	LD B,1FH	; B=number of values in list
0C92	DD 21 C0 0C		LD IX,LIST	; point index register X to 0CC0H
0C96	CB 85		RES 0,L	; reset exchange flag (bit 0 of L)
0C98	DD 5E 00	NEXT	LD E,(IX+0)	; get first number
0C9B	DD 7E 01		LD A,(IX+1)	; get next number on list
0C9E	BB		CP E	; compare the two numbers
0C9F	30 08		JR NC,NOEXCH	; if in order, do not exchange
0CA1	DD 73 01		LD (IX+1),E	; otherwise, exchange the contents
0CA4	DD 77 00		LD (IX+0),A	; of the two locations
0CA7	CB C5		SET 0,L	; set exchange flag (bit 0 of L)
0CA9	DD 23	NOEXCH	INC IX	; point X further up the list
0CAB	10 EB		DJNZ NEXT	; repeat until end of list
0CAD	CB 45		BIT 0,L	; test exchange flag
0CAF	20 DF		JR NZ,SORT	; repeat until exchange flag reset
0CB1	76		HALT	; program end

Problem 9 Two 8-bit numbers are stored in memory locations $0C80_{16}$ and $0C81_{16}$. Write a program (including algorithm and flowchart), using Z80 code, to **multiply** these two numbers together and store the product in memory locations $0C82_{16}$ and $0C83_{16}$.

Algorithm:

There are many different ways of multiplying numbers. One convenient method is to '**shift and add**' in a similar manner to that used for multiplying decimal numbers, and this method may be summarised as follows:

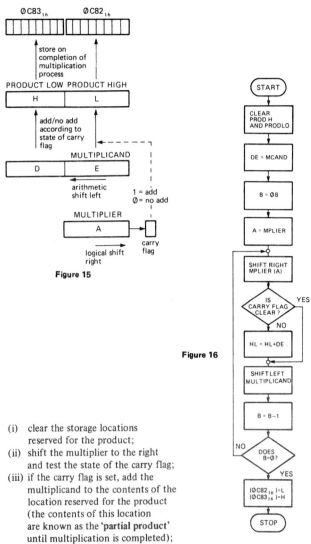

Figure 15

Figure 16

(i) clear the storage locations reserved for the product;

(ii) shift the multiplier to the right and test the state of the carry flag;

(iii) if the carry flag is set, add the multiplicand to the contents of the location reserved for the product (the contents of this location are known as the **'partial product'** until multiplication is completed);

(iv) if the carry flag is clear, add nothing to the partial product;

(v) shift the multiplicand to the left one place;

(vi) repeat (ii) to (v) a total of eight times;

(vii) stop.

Note, in the expression $X \times Y = Z$, X is called the **multiplicand**, Y is called the **multiplier** and Z is called the **product**.

This process is illustrated in *Figure 15*. Note, an 8-bit by 8-bit multiplication may give a product up to 16-bits, therefore two memory locations are reserved for storing the product.

Flowchart: see *Figure 16*.

Program code:

Machine code				Assembly language
	MCAND	EQU	0C80H	
	MPLIER	EQU	0C81H	
	PRODLO	EQU	0C82H	
	PRODHI	EQU	0C83H	
		ORG	0C90H	
0C90 21 00 00		LD HL,0		; clear product register
0C93 55		LD D,L		; clear high byte of multiplicand
0C94 3A 80 0C		LD A,(MCAND)		; get multiplicand into register A
0C97 5F		LD E,A		; and transfer to register E
0C98 06 08		LD B,08H		; B=loop counter (8 shift & add)
0C9A 3A 81 0C		LD A,(MPLIER)		; get multiplier into register A
0C9D CB 3F	SHIFT	SRL A		; shift multiplier right one place
0C9F 30 01		JR NC,NOADD		; add nothing if carry clear
0CA1 19		ADD HL,DE		; add multiplicand to partial ; product
0CA2 EB	NOADD	EX DE,HL		; temporarily swap registers
0CA3 29		ADD HL,HL		; equivalent to shift HL left
0CA4 EB		EX DE,HL		; swap registers back again
0CA5 10 F6		DJNZ SHIFT		; done 8 times? if not loop back
0CA7 22 82 0C		LD (PRODLO),HL		; store product in memory
0CAA 76		HALT		; program end

Problem 10 A block of 32_{10} numbers is located in memory, starting at address $0CC0_{16}$. Write a program (including algorithm and flowchart), using Z80 code, to find the **largest number** in the block and store this value in memory location $0C8F_{16}$.

Algorithm:

(i)   load register B with the length of the block;

(ii) load register L with the first number in the block;

(iii) load register E with numbers from the block using indexed addressing (index register X);

(iv) compare contents of register E with contents of register L (transfer L to accumulator to enable this to take place);

(v)   if the contents of register E are greater than the contents of register L, swap the contents of these registers;

(vi) increment index register X by 1 and decrement register B by 1;

(vii) repeat (iii) to (vi) until register B contains zero;

(viii) store the contents of register L in memory at location $0C8F_{16}$;
(ix) stop.

Flowchart: see *Figure 17.*

Figure 17

Program code:

Machine code		Assembly language	
	MAX	EQU 0C8FH	
	BLOCK	EQU 0CC0H	
		ORG 0C90H	
0C90 06 1F		LD B,1FH	; set up B as loop counter
0C92 DD 21 C1 0C		LD IX,BLOCK+1	; point X to second number in list
0C96 3A C0 0C		LD A (BLOCK)	; load register A with first on list
0C99 6F		LD L,A	; load register L with first on list
0C9A DD 5E 00	COMP	LD E,(IX+0)	; load register E with rest of block
0C9D BB		CP E,A	; compare E with A (and there- ; fore, L)
0C9E 30 02		JR NC,NEXT	; if L is larger, go to NEXT
0CA0 EB		EX DE,HL	; put larger number in L
0CA1 7D		LD A,L	; and transfer to A for next comp.
0CA2 DD 23	NEXT	INC IX	; point X to next number in block
0CA4 10 F4		DJNZ COMP	; repeat if not at end of block
0CA6 32 8F 0C		LD (MAX),A	; store maximum value in 0C8FH
0CA9 76		HALT	; program end

Problem 11 A list of non-zero numbers is stored in memory starting at location $0CC0_{16}$, and is terminated with 00_{16}. Write a program (including algorithm and flowchart), using Z80 code, to determine the **length** of the list and store this in memory location $0C8F_{16}$.

Algorithm:
(i) zero register pair HL (this counts the numbers in the list);
(ii) load register A (accumulator) with numbers from the list in succession, using index register X;
(iii) if register A contains a non-zero value, increment register pair HL, and repeat (ii) and (iii);
(iv) if register A contains zero, store HL in memory location $0C8F_{16}$;
(v) stop.

Flowchart: see *Figure 18*.

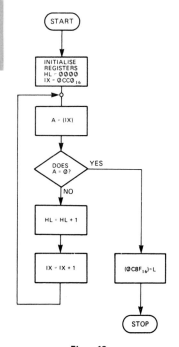

Figure 18

157

Program code:

Machine code		LENGTH	EQU 0C8FH	Assembly language
		LIST	EQU 0CC0H	
			ORG 0C90H	
0C90	21 00 00		LD HL,0	; clear HL register pair
0C93	DD 21 C0 0C		LD IX,LIST	; point index X to start of list
0C97	DD 7E 00	CHECK	LD A,(IX+0)	; get a number from the list
0C9A	B7		OR A	; set the flags
0C9B	28 05		JR Z,SAVE	; if zero, end of list reached
0C9D	23		INC HL	; increment number counter
				; (HL) by 1
0C9E	DD 23		INC IX	; point index X to next in list
0CA0	18 F5		JR CHECK	; not at end of list yet
0CA2	7D	SAVE	LD A,L	; transfer length to register A
0CA3	32 8F 0C		LD (LENGTH),A	; store length in location 0C8FH
0CA6	76		HALT	; program end

Problem 12 Two 8-bit numbers are stored in memory locations $0C80_{16}$ and $0C81_{16}$. Write a program (including algorithm and flowchart), using Z80 code, to **divide** the contents of $0C80_{16}$ by the contents of $0C81_{16}$ and store the quotient in location $0C82_{16}$ and any remainder in location $0C83_{16}$.

Algorithm:
(i)   clear registers A, C, E and H;
(ii)   load register L with the dividend from location $0C80_{16}$;
(iii)   load register D with the divisor from location $0C81_{16}$;
(iv)   using HL as a register pair, shift the dividend one place to the left,
(v)   subtract the divisor (register D) from register H;
(vi)   if the carry flag is set (indicating that D is greater than H), restore H register to its original value by adding D register to it;
(vii)   rotate the quotient in register C one place to the left, thus shifting in the current state of the carry flag;
(viii) carry out (iv) to (vii) a total of eight times;
(ix)   complement register C to invert the quotient (necessary due to the action of the carry flag when a borrow occurs);
(x)   store quotient (register C) in memory location $0C82_{16}$;
(xi)   store the remainder (register H) in memory location $0C83_{16}$;
(xii)   stop.
This process is illustrated in
Figure 19(a).

Flowchart: see *Figure 19(b).*

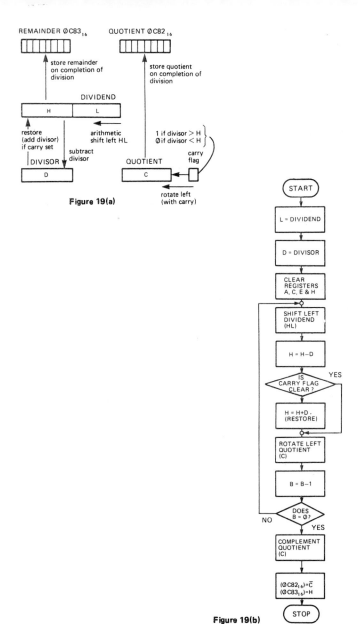

REMAINDER ∅C83₁₆   QUOTIENT ∅C82₁₆

store remainder
on completion of
division

store quotient
on completion of
division

DIVIDEND

H   L

restore
(add divisor)
if carry set

arithmetic
shift left HL

1 if divisor > H
∅ if divisor < H

DIVISOR

D

subtract
divisor

QUOTIENT

C

carry
flag

rotate left
(with carry)

Figure 19(a)

START

L = DIVIDEND

D = DIVISOR

CLEAR
REGISTERS
A, C, E & H

SHIFT LEFT
DIVIDEND
(HL)

H = H−D

IS
CARRY FLAG
CLEAR ?   YES

H = H+D .
(RESTORE)

ROTATE LEFT
QUOTIENT
(C)

B = B−1

DOES
B = ∅?

NO   YES

COMPLEMENT
QUOTIENT
(C)

(∅C82₁₆)=C̄
(∅C83₁₆)=H

STOP

Figure 19(b)

159

Program code:

Machine code	DIVDND	EQU 0C80H	Assembly language
	DIVSOR	EQU 0C81H	
	QUOTNT	EQU 0C82H	
	REMNDR	EQU 0C83H	
		ORG0C90H	
0C90 06 08		LD B,08H	; set up B as a loop counter
0C92 3A 80 0C		LD A,(DIVDND)	; load register A with dividend
0C95 6F		LD L,A	; and transfer to register L
0C96 3A 81 0C		LD A,(DIVSOR)	; load register A with divisor
0C99 57		LD D,A	; and transfer to register D
0C9A 97		SUB A	; clear register A
0C9B 67		LD H,A	; clear register H
0C9C 5F		LD E,A	; clear register E
0C9D 4F		LD C,A	; clear register C
0C9E 29	SHIFT	ADD HL,HL	; shift left dividend (in HL)
0C9F 7C		LD A,H	; temporarily move H into A
0CA0 92		SUB D	; subtract divisor from dividend
0CA1 30 01		JR NC,NOADD	; is divisor smaller than dividend
0CA3 82		ADD A,D	; if not, restore dividend
0CA4 CB 11	NOADD	RL C	; shift carry flag into C register
0CA6 67		LD H,A	; transfer A back into H
0CA7 10 F5		DJNZ SHIFT	; repeat if not done 8 times
0CA9 79		LD A,C	; transfer quotient (in C) to A
0CAA 2F		CPL	; complement quotient
0CAB 32 82 0C		LD(QUOTNT),A	; store quotient in location 0C82H
0CAE 7C		LD A,H	; transfer remainder (in H) to A
0CAF 32 83 0C		LD(REMNDR),A	; store remainder in location 0C83H
0CB2 76		HALT	; program end

Problem 13 A block of 32_{10} numbers is stored in memory starting at address 0080_{16}. Write a program (including algorithm and flowchart), using 6800 code, to **move** the block of numbers to a new starting address of $00B0_{16}$.

Algorithm:
(i) set up index register X as a loop counter by loading it with 0020_{16} (32_{10}),
(ii) load accumulator A with a number from the block, using indexed addressing;
(iii) store the contents of accumulator A in the new location, using indexed addressing;
(iv) decrement index register X by 1, and repeat (ii) and (iii) until X contains zero;
(v) stop.

Flowchart: see *Figure 20*.

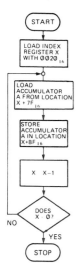

Figure 20

Program code:

Machine code			Assembly language
	BLOCK	EQU $0080	
	NEWLOC	EQU $00B0	
		ORG $0020	
0020 CE 00 20		LDX #$0020	; set up X as a loop counter
0023 A6 7F	LOOP	LDA A $7F,X	; select a number from the block
0025 A7 BF		STA A $BF,X	; move it to new location
0027 09		DEX	; point X to next in block
0028 26 F9		BNE LOOP	; repeat until whole block is moved
002A 3F		SWI	; program end

Note, this program copies the block of numbers into new locations and does not change the original block of numbers in any way.

Problem 14 A list of 32_{10} numbers is stored in memory, starting at address 0080_{16}. Write a program (including algorithm and flowchart), using 6800 code, to **sort** these numbers into ascending numerical order (all numbers are regarded as unsigned, i.e. all positive, for the purpose of this program).

Algorithm:
A version of the '**bubble sort**' described in *Worked Problem 2* may be used, and operates as follows:
(i) select the first number from the list and compare its magnitude with the second number in the list;
(ii) if the second number is larger than the first number (or the same), leave unchanged;

161

(iii) if the second number is smaller than the first, exchange the positions of the two numbers in the list and set an 'exchange flag' to indicate that an exchange takes place;

(iv) repeat (i) to (iii) with the second and third numbers in the list, third and fourth, etc., until the end of the list is reached;

(v) repeat (i) to (iv) until no further exchanges take place, determined by inspection of the 'exchange flag';

(vi) stop.

Flowchart: see *Figure 21*.

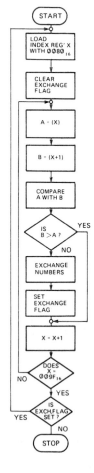

Figure 21

Program code:

Machine code	FLAG	EQU $007F	Assembly language
	LIST	EQU $0080	
		ORG $0020	
0020 CE 00 80	SORT	LDX #LIST	; point index reg. to start of list
0023 7F 00 7F		CLR FLAG	; clear exchange flag
0026 A6 00	NEXT	LDA A 0,X	; get first number in accumulator A
0028 E6 01		LDA B 1,X	; get next number in accumulator B
002A 11		CBA	; compare accumulators
002B 23 07		BLS NOEXCH	; if in order, do not exchange
002D A7 01		STA A 1,X	; otherwise, exchange the contents
002F E7 00		STA B 0,X	; of the two locations
0031 7C 00 7F		INC FLAG	; make exchange flag non-zero
0034 08	NOEXCH	INX	; point X further up the list
0035 8C 00 9F		CPX#LIST+$1F	; check for end of list
0038 26 EC		BNE NEXT	; repeat until end of list
003A 96 7F		LDA A FLAG	; get 'exchange flag'
003C 26 E2		BNE SORT	; keep sorting if not zero
003E 3F		SWI	; program end

Problem 15   Two 8-bit numbers are stored in memory locations 0080_{16} and 0081_{16}. Write a program (including algorithm and flowchart), using 6800 code, to **multiply** these two numbers together and store the product in memory locations 0082_{16} and 0083_{16}.

Algorithm: There are many different ways of multiplying numbers. One convenient method is to 'shift and add' in a similar manner to that used for multiplying decimal numbers, and this method may be summarised as follows:

(i)   clear the memory location reserved for the product;
(ii)   shift the multiplier to the right and test the state of the carry flag;
(iii)   if the carry flag is set, add the multiplicand to the contents of the memory location reserved for the product (the contents of this location are known as the 'partial product' until multiplication is completed);
(iv)   if the carry flag is clear, add nothing to the partial product;
(v)   shift the multiplicand to the left one place;
(vi)   repeat (ii) to (v) a total of eight times;
(vii) stop.

Note, in the expression $X \times Y = Z$, X is called the **multiplicand**, Y is called the **multiplier** and Z is called the **product**.

This process is illustrated in *Figure 22*. Note, the multiplicand is 8-bits wide and is shifted to the left a total of eight times. Therefore it is necessary to reserve a memory location to retain the bits of the multiplicand as they are shifted out of their initial location. Also, an

163

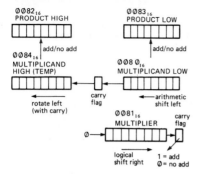

Figure 22

8-bit by 8-bit multiplication may give a product of up to 16 bits, therefore two memory locations are reserved for storing the product.

Flowchart: see *Figure 23*.

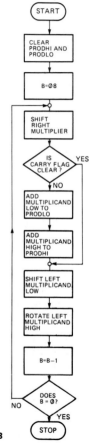

Figure 23

164

Program code:

Machine code				Assembly language
	MCAND	EQU	$0080	
	MPLIER	EQU	$0081	
	PRODHI	EQU	$0082	
	PRODLO	EQU	$0083	
	TEMP	EQU	$0084	
		ORG	$0020	
0020 5F		CLR B		; clear accumulator B
0021 D7 82		STA B PRODHI		; clear product high location
0023 D7 83		STA B PRODLO		; clear product low location
0025 D7 84		STA B TEMP		; clear temporary location
0027 C6 08		LDA B #$08		; set up accumulator B as a loop
				; counter
0029 74 00 81	SHIFT	LSR MPLIER		; shift multiplier right one place
002C 24 0C		BCC NOADD		; add nothing if carry clear
002E 96 83		LDA A PRODLO		; get partial product low byte
0030 9B 80		ADD A MCAND		; add on multiplicand low byte
0032 97 83		STA A PRODLO		; and put it back
0034 96 82		LDA A PRODHI		; get partial product high byte
0036 99 84		ADC A TEMP		; add on multiplicand high byte
0038 97 82		STA A PRODHI		; and put it back
003A 78 00 80	NOADD	ASL MCAND		; shift multiplicand left one place
003D 79 00 84		ROL TEMP		; rotate carry flag into TEMP
0040 5A		DEC B		; decrement accumulator B by 1
0041 26 E6		BNE SHIFT		; shifted 8 times? if not loop back
0043 3F		SWI		; program end

Problem 16 A block of 32_{10} numbers is located in memory, starting at address 0080_{16}. Write a program (including algorithm and flowchart), using 6800 code, to find the **largest number** in the block and store this value in memory location $00A0_{16}$.

Algorithm:

(i) load index register X with the length of the block;

(ii) load accumulator A with the last number in the block;

(iii) compare the remaining numbers in the block, in turn, with the number in accumulator A;

(iv) if a number larger than that which is in accumulator A is found, reload accumulator A with this value;

(v) decrement index register X by 1 after each comparison so that is is known when the first number in the block is reached and no further comparisons are necessary;

(vi) store accumulator A contents in memory at location $00A0_{16}$;

(vii) stop.

Flowchart: see *Figure 24.*

Program code:

Machine code		BLOCK	EQU $0080	Assembly language
		MAX	EQU $00A0	
			ORG $0020	
0020 CE 00 20			LDX #$0020	; load X with number of values ; in block
0023 96 9F			LDA A BLOCK+$1F	;load accumulator A with last in ; block
0025 A1 7F		COMP	CMP A BLOCK-1,X	; compare A with rest of block
0027 22 02			BHI NEXT	; if A holds larger number, go ; to NEXT
0029 A6 7F			LDA A BLOCK-1,X	; load A with new larger number
002B 09		NEXT	DEX	; point X to next number in block
002C 26 F7			BNE COMP	; repeat if not at end of block
002E 97 A0			STA A MAX	; store maximum value in $00A0
0030 3F			SWI	; program end

Problem 17 A list of non-zero numbers is stored in memory starting at location 0080_{16}, and is terminated with 00_{16}. Write a program (including algorithm and flowchart), using 6800 code, to determine the **length** of the list and store this in memory location $007E_{16}$.

Algorithm:
(i)   zero index register X;
(ii) load accumulator A with the first number in the list, using indexed addressing;
(iii) check the contents of accumulator A for zero value;
(iv) if accumulator A contains a non-zero value, increment index register X by 1 and repeat (ii) to (iv);
(v)   if accumulator A contains zero, store the value in index register X in memory location $007E_{16}$;
(vi) stop.

Flowchart: see *Figure 25*.

Program code:

Machine code		LENGTH	EQU $007E	Assembly language
		LIST	EQU $0080	
			ORG $0020	
0020 CE 00 00			LDX #$0000	; clear index register X
0023 A6 80		CHECK	LDA A $80,X	; get a number from the list
0025 27 03			BEQ SAVE	; if zero, end of list is reached
0027 08			INX	; increment number counter (X) ; by 1
0028 20 F9			BRA CHECK	; not at end of list yet
002A DF 7E		SAVE	STX LENGTH	; save list length in $007E
002C 3F			SWI	; program end

Problem 18 Two 8-bit numbers are stored in memory locations 0080_{16} and 0081_{16}. Write a program (including algorithm and flowchart), using 6800 code, to **divide** the contents of 0080_{16} by the contents of 0081_{16} and store the quotient in location 0082_{16} and any remainder in location 0083_{16}.

166

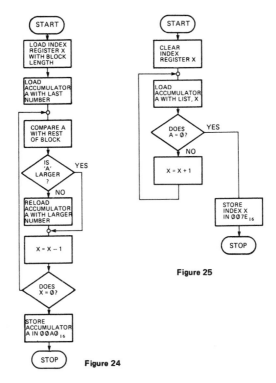

Figure 25

Figure 24

Algorithm:

(i) clear accumulators A and B;

(ii) shift the dividend one place to the left and shift any carry into accumulator A;

(iii) subtract the divisor from accumulator A;

(iv) if the carry flag is set (divisor larger than accumulator A), restore accumulator A by adding the divisor to it;

(v) rotate accumulator B (the quotient) to the left, thus shifting in the current state of the carry flag;

(vi) carry out (ii) to (v) a total of eight times;

(vii) complement accumulator B to invert the quotient (necessary due to the action of the carry flag when a borrow occurs);

(viii) store accumulator B (quotient) in memory location 0082_{16};

(ix) store accumulator A (remainder) in memory location 0083_{16};

(x) stop.

This process is illustrated in *Figure 26(a)*.

Note, in the expression $X \div Y = Z$, X is called the **dividend**, Y is called the **divisor** and Z is called the **quotient**.

Flowchart: see *Figure 26(b)*.

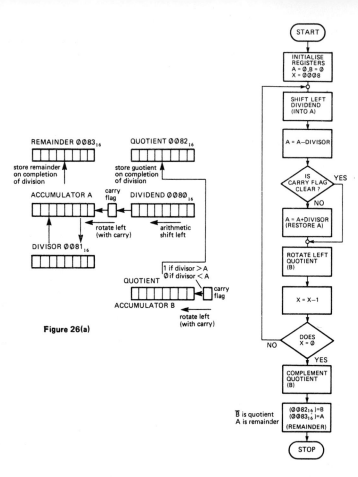

REMAINDER 0083₁₆ QUOTIENT 0082₁₆

store remainder on completion of division

store quotient on completion of division

ACCUMULATOR A

carry flag

DIVIDEND 0080₁₆

rotate left (with carry)

arithmetic shift left

DIVISOR 0081₁₆

1 if divisor > A
0 if divisor < A

QUOTIENT

carry flag

ACCUMULATOR B

rotate left (with carry)

Figure 26(a)

START

INITIALISE REGISTERS
A = 0, B = 0
X = 0008

SHIFT LEFT DIVIDEND (INTO A)

A = A − DIVISOR

IS CARRY FLAG CLEAR? — YES

NO

A = A + DIVISOR (RESTORE A)

ROTATE LEFT QUOTIENT (B)

X = X − 1

DOES X = 0

NO

YES

COMPLEMENT QUOTIENT (B)

(0082₁₆) = B
(0083₁₆) = A
(REMAINDER)

B̄ is quotient
A is remainder

STOP

Figure 26(b)

168

Program code:

Machine code	DIVDND	EQU $0080	Assembly language
	DIVSOR	EQU $0081	
	QUOTNT	EQU $0082	
	REMNDR	EQU $0083	
		ORG $0020	
0020 5F		CLR B	; clear accumulator B (quotient)
0021 4F		CLR A	; clear accumulator A
0022 CE 00 08		LDX #$0008	; set up X as a loop counter
0025 78 00 80	SHIFT	ASL DIVDND	; shift left DIVDND and accumu- ; lator A
0028 49		ROL A	; as a 16-bit register
0029 90 81		SUB A DIVSOR	; subtract DIVSOR from accumu- ; lator A
002B 24 03		BCC CLEAR	; is DIVSOR greater than accumu- ; lator A?
002D 9B 81		ADD A DIVSOR	; if so, restore accumulator A
002F 0D		SEC	; and restore state of carry flag
0030 59	CLEAR	ROL QUOTNT	; and shift a digit into the quotient
0031 09		DEX	; decrement X by 1
0032 26 F1		BNE SHIFT	; repeat if not done 8 times
0034 53		COM B	; complement quotient
0035 D7 82		STA B QUOTNT	; save quotient in location $0082
0037 97 83		STA A REMNDR	; save remainder in location $0083
0039 3F		SWI	; program end

C FURTHER PROBLEMS ON PROGRAMS WITH LOOPS

(a) SHORT ANSWER PROBLEMS

All problems in this section refer to Figure 27.

1 The technique illustrated by the flowchart in *Figure 27* is called . . . .

2 A register (or memory location), used to determine the number of times that the part of a program represented by block 'B' is executed, is called a . . . .

3 Block 'A' represents an . . . . section for the program.

4 Block 'B' represents the . . . .

5 The function most likely to be performed in block 'C' is to . . . . a register (or memory location).

6 Together, blocks 'C' and 'D' form a . . . . section.

7 Frequently, the operand of an instruction must be changed on each pass through block 'B'. For this purpose, ... addressing may be used.

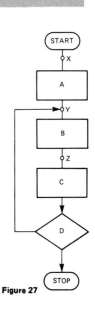

Figure 27

169

8   The operand of the instruction in block 'D' must be such as to allow the program counter to . . . .

9   If a bug causes the program to flow from block 'D' to point 'X' (instead of point 'Y'), the effect is to cause the program to . . . .

10 If a bug causes the program to flow from block 'D' to point 'Z' (instead of point 'Y'), the effect is to cause the program to . . . .

(b) CONVENTIONAL PROBLEMS

1   A block of 32_{10} numbers is located in memory, starting at address 0080_{16}. Write a program (including algorithm and flowchart), using 6502 code, to find the **smallest number** in the block and store this value in memory location $00A0_{16}$.

2   Write a program (including algorithm and flowchart), using 6502 code, to **clear** a block of memory (store 00 in each location) between 0080_{16} and $009F_{16}$ inclusive.

3   Write a program (including algorithm and flowchart), using 6502 code, to **reverse** the order of bits in memory location 0080_{16} (i.e. exchange bits 7 and 0, 6 and 1, 5 and 2, 4 and 3).

4   A list of 32_{10} numbers is stored in memory, starting at address 0080_{16}. Write a program (including algorithm and flowchart), using 6502 code, to **sort** these numbers into **descending** numerical order (all numbers are regarded as unsigned, i.e. all positive, for the purpose of this program).

5   A list of 32_{10} numbers is stored in memory, starting at address 0080_{16}. Write a program (including algorithm and flowchart), using 6502 code, to **count** the number of times that the data byte 00 occurs in this list.

6   Five decimal numbers are stored in memory locations 0080_{16} to 0084_{16} inclusive. Write a program (including algorithm and flowchart), using 6502 code, to **add** these numbers together, and store their decimal sum in memory locations 0085_{16} (hundreds) and 0086_{16} (tens and units).

7   A block of 32_{10} numbers is located in memory, starting at address $0CC0_{16}$. Write a program (including algorithm and flowchart), using Z80 code, to find the **smallest number** in the block and store this value in memory location $0CBF_{16}$.

8   Write a program (including algorithm and flowchart), using Z80 code, to **clear** a block of memory (store 00 in each location) between $0CC0_{16}$ and $0CDF_{16}$ inclusive.

9   Write a program (including algorithm and flowchart), using Z80 code, to **reverse** the order of bits in memory location $0C80_{16}$ (i.e. exchange bits 7 and 0, 6 and 1, 5 and 2, 4 and 3).

10 A list of 32_{10} numbers is stored in memory, starting at address $0CC0_{16}$. Write a program (including algorithm and flowchart), using Z80 code, to **sort** these numbers into descending numerical order (all numbers are regarded as unsigned, i.e. all positive, for the purpose of this program).

11 A list of 32_{10} numbers is stored in memory, starting at address $0CC0_{16}$. Write a program (including algorithm and flowchart), using Z80 code, to **count** the number of times that the data byte 00 occurs in this list.

170

12 Five decimal numbers are stored in memory locations $\emptyset C8\emptyset_{16}$ to $\emptyset C84_{16}$ inclusive. Write a program (including algorithm and flowchart), using Z80 code, to **add** these numbers together, and store their decimal sum in memory locations $\emptyset C85_{16}$ (hundreds) and $\emptyset C86_{16}$ (tens and units).

13 Repeat *Problem 1* using 6800 code.

14 Repeat *Problem 2* using 6800 code.

15 Repeat *Problem 3* using 6800 code.

16 Repeat *Problem 4* using 6800 code.

17 Repeat *Problem 5* using 6800 code.

18 Repeat *Problem 6* using 6800 code.

6 Interfacing

A MAIN POINTS CONCERNED WITH INTERFACING

1 In order that a microcomputer may interact with external devices and thus form part of a control system, a means of getting data into and out of it must be provided. External devices attached to a microcomputer for this purpose are called '**peripheral devices**' or '**peripherals**', and examples of these include keyboards, transducers of various types, light-emitting diodes (LEDs), visual display units (VDUs) and printers.

2 It is not normally possible to connect peripheral devices directly to a microcomputer bus system due to a lack of compatibility. To enable two otherwise incompatible systems to be interconnected, an electronic circuit known as an '**interface**' must be used. The action of using an interface circuit is known as '**interfacing**'.

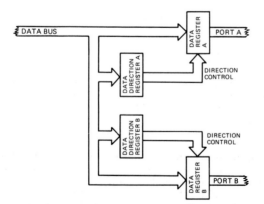

Figure 1(a)

Figure 1(b)

3   It is not possible to connect a peripheral device directly to a microcomputer data bus, since data on this bus changes continuously due to normal microprocessor activity. Therefore, data to be transferred to a peripheral device are clocked into a set of data latches (flip-flops) during memory-write cycles, i.e. during 'store' instructions. Latches retain their data, i.e. act as small memories, thus giving peripheral devices time in which to read the data.

     Input peripherals are connected to a microcomputer data bus via tri-state buffers. When selected, tri-state buffers enable a peripheral device to put data onto a microcomputer data bus. When deselected, tri-state buffers isolate a peripheral device from a microcomputer data bus and prevent it from disturbing normal data bus activity.

4   The functions described in paragraph 3 are frequently provided in special purpose programmable interface chips called 'peripheral interface adapters' (PIA) or 'parallel input–output' (PIO) devices. A simplified block diagram of a typical PIA is illustrated in *Figure 1(a)*.

     Two 8-bit ports are provided on most PIAs, generally referred to as 'Port A' and 'Port B'. Each of the 8 bits of a specific port is defined as an input or an output according to the contents of the port's 'data direction register'. The diagram shows that a logical $\emptyset$ in a particular bit position of the data direction register defines the corresponding port bit as an input, whilst a logical 1 defines the corresponding port bit as an output (see *Figure 1(b)*).

5   As an example of interfacing with a PIA, consider the connection of four single pole single throw (SPST) switches to Port B, and eight light-emitting diodes (LEDs) to Port A. The necessary interfacing components are illustrated in *Figure 2*.

     The data buffers on a PIA are only capable of delivery of a very small current, but LEDs require anything up to 20 mA for full brilliance. Therefore, electrical buffering is required between the PIA and the LEDs, and takes the form of two 7401 logic ICs which are used in this example as current amplifiers. The 270 Ω resistors, R1–R8, act as current limiters to prevent excessive current through the LEDs when illuminated.

     Four SPST switches are connected to $b_\emptyset$–b_3 of Port B to allow any hex. value $\emptyset$–F to be set up at its inputs. Resistors R9–R12 act as 'pull-up' resistors to hold each input to Port B at logical 1 until its particular input switch is closed, when an input of logical $\emptyset$ is obtained. Port B b_4–b_7 are unused in this application and must be masked off by appropriate software, or electrically connected to 0 V (logical $\emptyset$).

173

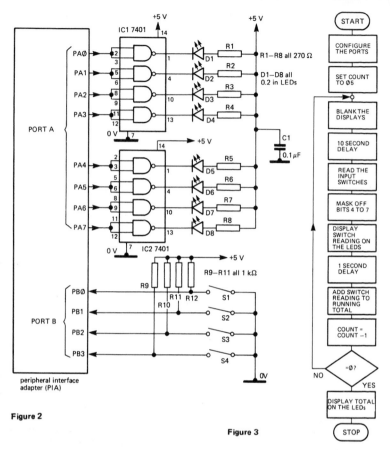

Figure 2

Figure 3

6   As an example of a simple application for the interface circuit shown in *Figure 2*, the addition program described in Chapter 5, paragraph 5, may be modified so that five numbers are obtained from input switches S–S4 (in binary form), and their sum displayed on the LEDs (in binary form). Program modifications are as follows:

(i)   incorporate a time delay of approximately 10 s to allow time for numbers to be set up on S1–S4,

(ii)   read Port B for each of the five numbers instead of obtaining them from a table in memory.

(iii)   display the value of the switches on the LEDs for approximately one second to indicate that the switches have been read and thus prompt the user to enter the next number;

(iv)   display the computed sum on the LEDs.

A flowchart for the modified program is given in *Figure 3* and program code is as follows:

(a) 6502 code

Machine code			SUM=$0080 DELAY1=$0081 DELAY2=$0082 DELAY3=$0083 PAD=$1700 PADD=$1701 *=$0020	Assembly language	
0020	A9	FF		LDA#$FF	
0022	8D	01 17		STA PADD	; configure Port A as all outputs
0025	A9	00		LDA#$00	
0027	85	80		STA SUM	; clear sum location
0029	8D	03 17		STA PBDD	; configure Port B as all inputs
002C	A2	05		LDX#$05	; X=number of values to add
002E	A9	00	BLANK	LDA#$00	
0030	8D	00 17		STA PAD	; blank out the LEDs
0033	A9	13		LDA# 13	; set up delay time
0035	85	83		STA DELAY3	
0037	C6	81	LOOP1	DEC DELAY1	; 10 second delay
0039	D0	FC		BNE LOOP1	
003B	C6	82		DEC DELAY2	
003D	D0	F8		BNE LOOP1	
003F	C6	83		DEC DELAY3	
0041	D0	F4		BNE LOOP1	
0043	AD	02 17		LDA PBD	; read the switches
0046	29	0F		AND#$0F	; mask off bits 4 to 7
0048	8D	00 17		STA PAD	; and light up the LEDs
004B	C6	81	LOOP2	DEC DELAY1	; 0.5 second delay
004D	D0	FC		BNE LOOP2	
004F	C6	82		DEC DELAY2	
0051	D0	F8		BNE LOOP2	
0053	18			CLC	; clear carry for addition
0054	65	80		ADC SUM	; add running total to accumulator
0056	85	80		STA SUM	; and put new total back
0058	CA			DEX	; count five numbers
0059	D0	D3		BNE BLANK	; get another if not the last
005B	8D	00 17		STA PAD	; display total
005E	00			BRK	; program end

Note: This program is written to run on a 6502 based microcomputer which operates with a 1 MHz clock, and I/O addresses as follows:

Port A Data register	$1700	Port B Data register	$1702
Port A Data Direction register	$1701	Port B Data Direction register	$1703

Changes may be necessary to the delay constant at address $0034, and the I/O addresses if equipment different to that used by the author is substituted.

(b) Z80 code

Machine code				Assembly language
		ORG 0C90H		
0C90	3E 0F		LD A,0FH	
0C92	D3 06		OUT (6),A	; PIO side A all outputs
0C94	3E 4F		LD A,4FH	
0C96	D3 07		OUT (7),A	; PIO side B all inputs
0C98	06 05		LD B,05H	; B=number of values to add
0C9A	AF		XOR A	; clear register A
0C9B	4F		LD C,A	; clear register C
0C9C	D3 04	BLANK	OUT (4),A	; blank out the LEDs
0C9E	21 00 08		LD HL,0800H	; set up delay time
0CA1	1C	LOOP1	INC E	; 10 second delay
0CA2	20 FD		JR NZ,LOOP1	
0CA4	2B		DEC HL	
0CA5	7D		LD A,L	
0CA6	B4		OR H	; set Z flag when HL=0
0CA7	20 F8		JR NZ,LOOP1	
0CA9	DB 05		IN A,(5)	; read the switches
0CAB	E6 0F		AND 0FH	; mask off bits 4 to 7
0CAD	D3 04		OUT (4),A	; and light up the LEDs
0CAF	81		ADD A,C	; add running total to A
0CB0	4F		LD C,A	; and put new total back in C
0CB1	21 00 02		LD HL,0200H	; 1 second delay
0CB4	1C	LOOP2	INC E	
0CB5	20 FD		JR NZ,LOOP2	
0CB7	2B		DEC HL	
0CB8	7D		LD A,L	
0CB9	B4		OR H	; set Z flag when HL=0
0CBA	20 F8		JR NZ,LOOP2	
0CBC	AF		XOR A	; clear register A
0CBD	10 DD		DJNZ BLANK	; get next value if not last
0CBF	79		LD A,C	; get total into A
0CC0	D3 04		OUT (4),A	; display total
0CC2	76		HALT	; program end

Note: This program is written to run on a Z80 based microcomputer which operates with a 2 MHz clock, and I/O ports as follows:

PIO side A data	Port 4	PIO side B data	Port 5
PIO side A control	Port 6	PIO side B control	Port 7

Changes may be necessary to the delay constants at addresses 0CA0H and 0CB3H, and the port numbers in I/O instructions if equipment different to that used by the author is substituted.

(c) 6800 code

Machine code			Assembly language
	SUM EQU $0080		
	DRA EQU $8004		
	CRA EQU $8005		
	DRB EQU $8006		
	CRB EQU $8007		
	ORG $0020		
0020 4F		CLR A	
0021 B7 80 05		STA A CRA	; select direction register A
0024 B7 80 07		STA A CRB	; select direction register B
0027 97 80		STA A SUM	; clear sum location
0029 B7 80 06		STA A DRB	; Port B all inputs
002C 86 FF		LDA A#$FF	
002E B7 80 04		STA A DRA	; Port A all outputs
0031 86 04		LDA A#$04	
0033 B7 80 05		STA CRA	; select I/O register A
0036 B7 80 07		STA CRB	; select I/O register B
0039 C6 05		LDA B#$05	; B=number of values to add
003B 4F	BLANK	CLR A	
003C B7 80 04		STA A DRA	; blank out the LEDs
003F 86 05		LDA A #$05	; set up delay time
0041 09	LOOP1	DEX	; 10 second delay
0042 26 FD		BNE LOOP1	
0044 4A		DEC A	
0045 26 FA		BNE LOOP1	
0047 B6 80 06		LDA A DRB	; read the switches
004A 84 0F		AND A #$0F	; mask off bits 4 to 7
004C B7 80 04		STA A DRA	; and light up the LEDs
004F 09	LOOP2	DEX	; 1 second delay
0050 26 FD		BNE LOOP2	
0052 9B 80		ADD A SUM	; add running total to accumulator
0054 97 80		STA A SUM	; and put new total back
0056 5A		DEC B	; count five numbers
0057 26 E2		BNE BLANK	; get another if not the last
0059 B7 80 04		STA A DRA	; display total
005C 3F		SWI	; program end

Note; This program was written to run on a 6800 based microcomputer which operates with a 614 kHz clock, and I/O addresses as follows:

Port A I/O register	$8004	Port B I/O register	$8006
Port A Data Direction register	$8004	Port B Data Direction register	$8006
Port A Control register	$8005	Port B Control register	$8007

Changes may be necessary to the delay constant at address $0040, and the I/O addresses if equipment different to that used by the author is substituted.

Problem 1   With reference to the block diagram of a 6530 peripheral interface adapter (PIA) shown in *Figure 3*,
(a) explain what is meant by '**configuring**' a PIA, and
(b) explain how this PIA may be configured according to *Table 1* (use 6502 code and any convenient PIA addresses).

(a) A peripheral interface adapter (PIA) provides a versatile programmable interface between a microcomputer and its peripheral devices. Many options may be open to the user of a particular PIA, and, determining which options are to be used, and how they are to be used is the process of '**configuring**' or '**initialising**' a PIA.

(b) One essential task when configuring a PIA is to determine which bits of each port are to be outputs and which bits are to be inputs. For this purpose, a 6530 PIA has two '**data control**' or '**data direction**' registers, one for Port A and the other for Port B. For programming purposes, these registers may be considered as two individual memory locations which may be treated in the same way as, for example, random access memory (RAM). Where a logical 1

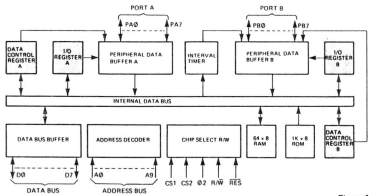

Figure 4

is stored in a particular bit position in a data direction register, its corresponding port I/O bit is programmed as an **output**. Where a logical $\emptyset$ is stored in a particular bit position, its corresponding port I/O bit is programmed as an **input**.

A 6530 PIA also has two I/O or data registers (one for each port), whose function is to capture and store (latch) data from the data bus which are destined as outputs to peripheral devices. These four registers (shaded on *Figure 4*) are '**memory mapped**' and occupy four contiguous addresses which for this example may be considered to be as follows:

Table 1

Port	bit7	bit6	bit5	bit4	bit3	bit2	bit1	bit∅
A	input	output	output	output	input	input	input	input
B	input	output	output	input	input	input	input	input

Port A Data (I/O) register	(PAD)	$1700
Port A Data Direction (control) register	(PADD)	$1701
Port B Data (I/O) register	(PBD)	$1702
Port B Data Direction (control) register	(PBDD)	$1703

To configure the PIA according to *Table 1*, using these addresses, the following program segment is required:

```
A9  70           LDA #%01110000
8D  01  17       STA PADD          ; configure Port A
A9  60           LDA #%01100000
8D  03  17       STA PBDD          ; configure Port B
```

(*Note:* % = a binary number)

Problem 2   With reference to the block diagram of Z80 parallel input/output (PIO) chip shown in *Figure 5(a) and (b)*,
(i)   explain how this device is addressed;
(ii) explain the function of the '**mode control register**';
(iii) assuming Port A and Port B use I/O addresses specified in *Table 2*, show how a PIO of this type may be configured so that:
(a)   all of Port A I/O lines are outputs with bits ∅–5 at logical ∅ and bits 6 and 7 at logical 1;
(b)   all of Port B I/O lines are inputs and data from a peripheral device are read into register A (accumulator);
(c)   Port A is used in the control mode with bits 2, 3 and 4 as inputs, and the remainder of the bits as outputs.

(i)   The Z80 microprocessor has special instructions for input/output operations, therefore the PIO designed for use with this microprocessor is of the '**isolated I/O**' type, i.e. it is not memory mapped, and addresses are decoded separately for memory and I/O.

   During execution of IN and OUT instructions, the second byte of the instruction provides a port address which appears on the microprocessor address bus lines A∅ to A7. Selection of a particular I/O facility is obtained by means of three signals:
(a) $\overline{IORQ}$, which becomes logical ∅ during execution of IN and OUT instructions;

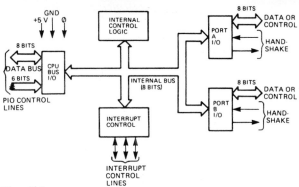

Figure 5(a)

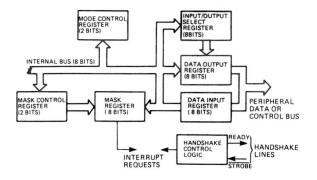

Figure 5(b)

    (b) B/A SEL, which selects Port A of a PIO if at logical $\emptyset$, otherwise Port B is selected,

    (c) CONTROL/DATA SEL, which selects the PIO data I/O registers if at logical $\emptyset$, otherwise the control registers are selected.

One possible method of connecting a single Z80 PIO to a Z80 microprocessor is shown in *Figure 6*.

  (ii) A Z80 PIO is capable of operating in **four** different modes:

    (a) byte output mode; (b) byte input mode; (c) bi-directional mode; and (d) bit mode.

The '**mode control register**' is a two bit register which is loaded by a microprocessor to select one of the four modes available. A PIO recognises a 'write to mode control register' request as a data word with bits $\emptyset$–3 all at logical 1 (see *Figure 7*). The four operating modes are summarised in *Table 3*.

180

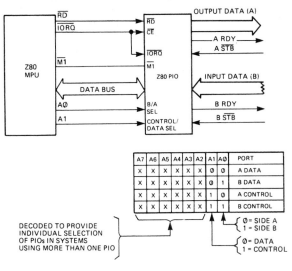

Figure 6

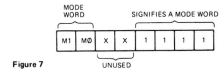

Figure 7

(iii) Using Z80 code, a PIO may be configured as follows:

(a) All of Port A I/O lines as outputs with bits Ø–5 at logical Ø, and bits 6 and 7 at logical 1:

```
3E ØF    LD A,ØFH
D3 06    OUT (6),A      ; Port A byte output mode
3E CØ    LD A,CØH
D3 04    OUT (4),A      ; set bits 6 & 7, reset bits Ø–5
```

Table 2

Port	I/O register address	Control register address
A	4	6
B	5	7

Table 3

Mode	M1	MØ	Operating mode
Ø	Ø	Ø	**Byte output**: all bits of a port are defined as outputs, and data may be written to a peripheral device by the microprocessor.
1	Ø	1	**Byte input**: all bits of a port are defined as inputs, and data may be read from a peripheral device by the microprocessor.
2	1	Ø	**Bi-directional**: a port may be read or written to, but this facility is only available on port A since all four handshake lines are used.
3	1	1	**Bit mode**: this mode is intended for status and control applications. When this mode is selected, the next word must set the I/O register to define which lines are inputs and which are outputs (Ø=system output and 1=system input).

(b) All of Port B I/O lines as inputs, with data read from peripheral device stored in register A (accumulator):

```
3E 4F     LD A,4FH
D3 Ø7     OUT (7),A     ; Port B byte input mode
DB Ø5     IN A,(5)      ; read Port B into A
```

(c) Port A in the control mode with bits 2, 3 and 4 as inputs and the remainder as outputs:

```
3E CF     LD A,CFH
D3 Ø6     OUT (6),A     ; Port A in bit mode
3E 1C     LD A,1CH
D3 Ø6     OUT (6),A     ; define inputs and outputs
                        ; (1=input, Ø=output)
```

Problem 3 With reference to the block diagram of a 6820 Peripheral Interface Adapter (PIA) shown in *Figure 8*:
(i) explain how it is possible to select each of the six user-accessible registers (shown shaded) with only two register-select lines (RSØ and RS1).
(ii) show how a 6820 PIA may be configured according to *Table 4*.

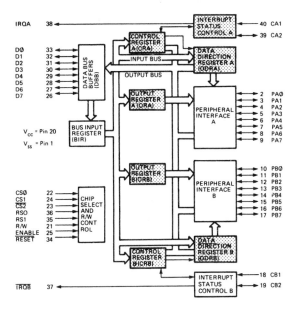

IRQA 38

DO 33
D1 32
D2 31
D3 30
D4 29
D5 28
D6 27
D7 26

V_{cc} = Pin 20
V_{ss} = Pin 1

CSØ 22
CS1 24
CS2 23
RS0 36
RS1 35
R/W 21
ENABLE 25
RESET 34

IRQB 37

Figure 8

(i) The same address is used to access either a port I/O register or its data direction register. Which register is actually accessed on application of this address is determined by bit 2 of the port's control register, as illustrated in *Figure 9*. This shows:

(a) if bit 2 of a control register is reset (logical Ø), a data direction register is selected; and

(b) if bit 2 of a control register is set (logical 1), a data I/O register is selected. Thus, all user accessible registers in a 6820 PIA are contained in four contiguous memory locations. Once reset, e.g. after switching on a microcomputer, bit 2 in each control register of a 6820 PIA is reset to Ø, therefore the data direction registers are selected ready for port configuration.

(ii) For the purposes of this problem, addresses for the PIA are assumed to be as follows:

Port A I/O register (ORA)	$8004
Port A Data Direction register (DDRA)	$8004
Port A Control register (CRA)	$8005
Port B I/O register (ORB)	$8006
Port B Data Direction register (DDRB)	$8006
Port B Control register (CRB)	$8007

Using these addresses, configuration of a 6820 PIA according to *Table 4* may be accomplished with the following program segment:

4F	CLRA	SELECT DATA DIRECTION REGISTERS (NOT NECESSARY EXCEPT WHEN RECONFIGURING A PIA DURING A PROGRAM RUN)
B7 80 05	STA A CRA	
B7 80 07	STA A CRB	
86 E3	LDA A #%11100011	DEFINE PORT A & B INPUTS AND OUTPUTS IN DATA DIRECTION REGISTERS
B7 80 04	STA A DDRA	
86 19	LDA A #%00011001	
B7 80 06	STA A DDRB	
86 04	LDA A #$04	SET BIT2 ON CONTROL REGISTERS AND THUS SELECT I/O REGISTERS
B7 80 05	STA A CRA	
B7 80 07	STA A CRB	

Table 4

Port	bit7	bit6	bit5	bit4	bit3	bit2	bit1	bit0
A	output	output	output	input	input	input	output	output
B	input	input	input	output	output	input	input	output

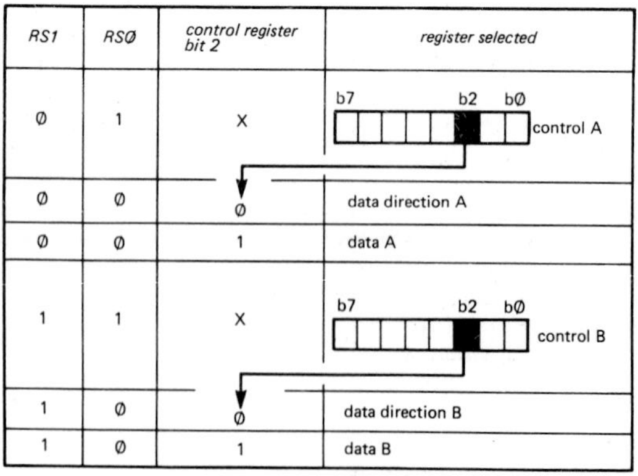

Figure 9

A single, common anode, 7-segment LED may be interfaced to a microcomputer by the use of a circuit similar to that shown in *Figure 10*.

Each segment of an LED display requires a current of 10–20 mA when illuminated. A PIA cannot deliver a current of this magnitude. Therefore a buffer (current amplifier) must be used, and one possible circuit (*Figure 10*) makes use of logic gates connected as current amplifiers. A resistor array (seven 150 Ω resistors) is used to limit current flow through illuminated segments to a safe value.

One main difficulty which arises when operating 7-segment displays concerns the lack of any obvious relationship between a number (or letter) and the segment pattern required to form it. A hardware solution takes the form of 'BCD to 7-segment' decoder chips, but these only allow numbers 0 to 9 to be displayed and are therefore unsuitable for many applications.

A software solution allows greater flexibility, and takes the form of a **segment code table** (called a 'look-up' table), which is stored in memory, and accessed by using the number or letter to be displayed as an index. This system is illustrated in *Figure 11*.

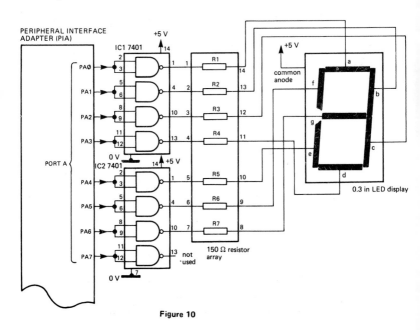

Figure 10

185

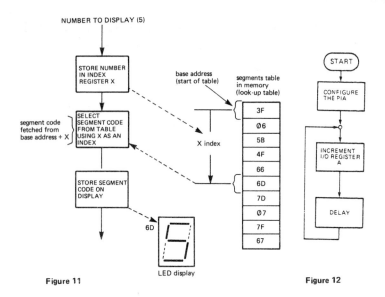

NUMBER TO DISPLAY (5)

STORE NUMBER IN INDEX REGISTER X

segment code fetched from base address + X

SELECT SEGMENT CODE FROM TABLE USING X AS AN INDEX

STORE SEGMENT CODE ON DISPLAY

base address (start of table)

segments table in memory (look-up table)

| 3F |
| 06 |
| 5B |
| 4F |
| 66 |
| 6D |
| 7D |
| 07 |
| 7F |
| 67 |

X index

6D

LED display

Figure 11

START

CONFIGURE THE PIA

INCREMENT I/O REGISTER A

DELAY

Figure 12

Problem 5 A 6502 based microcomputer with a 6530 PIA is connected to eight LEDs using an interface circuit of the type illustrated in *Figure 2* of this chapter. Assuming the PIA Port A data and data direction registers are located at addresses $1700 and $1701 respectively, write a program to make the display count up slowly in binary. Include a flowchart in your answer.

Flowchart: see *Figure 12*.

Program code:

Machine code		PAD=$1700	Assembly language
		PADD=$1701	
		*=$0020	
0020 A9 FF		LDA #$FF	
0022 8D 01 17		STA PADD	; Port A all outputs
0025 EE 00 17	COUNT	INC PAD	; increment count by 1
0028 CA	DELAY	DEX	; time delay
0029 D0 FD		BNE DELAY	
002B 88		DEY	
002C D0 FA		BNE DELAY	
002E F0 F5		BEQ COUNT	

186

Problem 6 A 6502 based microcomputer with a 6530 PIA is connected to four SPST switches and a 7-segment LED display using the interface circuits shown in *Figures 2 and 10*. Assuming Port A data and direction registers are located in addresses $1700 and $1701 respectively, and Port B data and direction registers are located at addresses $1702 and $1703 respectively, write a program in 6502 code such that the binary value selected on the switches is displayed as a hexadecimal character on the 7-segment display. Include a flowchart in your answer.

Flowchart: see *Figure 13*.

Program code

Machine code			PAD=$1700 PADD=$1701 PBD=$1702 PBDD=$1703 TABLE=$0037 *=$0020	Assembly language
0020	A9	7F	LDA#$7F	
0022	8D 01	17	STA PADD	; configure Port A
0025	A9	00	LDA $00	
0027	8D 03	17	STA PBDD	; configure Port B
002A	AD 02	17	READ LDA PBD	; read the switches
002D	29	0F	AND #$0F	; mask off bits 4 to 7
002F	AA		TAX	; use switch reading as index
0030	B5	37	LDA TABLE,X	; get segments code from table
0032	8D 00	17	STA PAD	; light up display
0035	D0	F3	BNE READ	; =branch always to $002A
0037	3F		TABLE .BYTE $3F,$06,$5B; 0, 1, 2	
0038	06			
0039	5B			
003A	4F		.BYTE $4F,$66,$6D ; 3, 4, 5	
003B	66			
003C	6D			
003D	7D		.BYTE $7D,$07,$7F ; 6, 7, 8	
003E	07			
003F	7F			
0040	67		.BYTE $67,$77,$7C ; 9, A, B	
0041	77			
0042	7C			
0043	39		.BYTE $39,$5E,$79 ; C, D, E	
0044	5E			
0045	79			
0046	71		.BYTE $71 ; F	

Problem 7 A 6502 based microcomputer, with a 6530 PIA, is connected to a single 7-segment display using the interface circuit shown in *Figure 10* of this chapter. Assuming Port A data and data direction registers are located at addresses $1700 and $1701 respectively, write a program in 6502 code to display a count of 0 to 9 repetitively at one second intervals. Include a flow-chart in your answer.

Flowchart: see *Figure 14*.

Program code:

Machine code			DELAY1=$0080	Assembly language
			DELAY2=$0081	
			PAD=$1700	
			PADD=$1701	
			*=$0020	
0020	A9 FF		LDA#$FF	
0022	8D 01	17	STA PADD	; Port A all outputs
0025	A2 09		RESET LDX#$09	; X points to end of table
0027	B5 3E		COUNT LDA TABLE,X	; get segments code
0029	8D 00	17	STA PAD	; light up display
002C	A0 10		LDY#$10	; delay constant
002E	C6 80		LOOP DEC DELAY1	; 1 second delay
0030	D0 FC		BNE LOOP	
0032	C6 81		DEC DELAY2	
0034	D0 F8		BNE LOOP	
0036	88		DEY	
0037	D0 F5		BNE LOOP	
0039	CA		DEX	; next segment code
003A	10 EB		BPL COUNT	; carry on counting if positive
003C	30 E7		BMI RESET	; start from zero again
003E	67		TABLE .BYTE $67,$7F,$07 ; 9, 8, 7	
003F	7F			
0040	07			
0041	7D		.BYTE $7D,$6D,$66 ; 6, 5, 4	
0042	6D			
0043	66			
0044	4F		.BYTE $4F,$5B,$06 ; 3, 2, 1	
0045	5B			
0046	06			
0047	3F		.BYTE $3F ; 0	

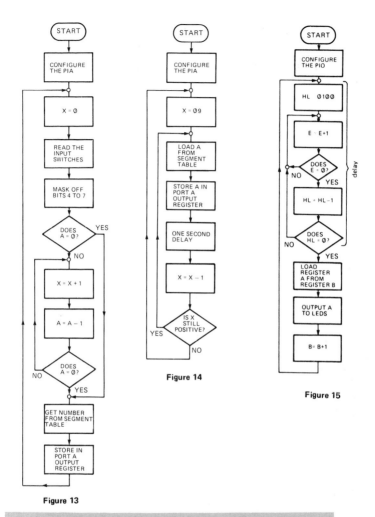

Figure 13

Figure 14

Figure 15

Problem 8 A Z80 based microcomputer with a Z80 P10 is connected to eight LEDs using an interface circuit of the type illustrated in *Figure 2* of this chapter. Assuming Port A data and mode control registers occupy I/O addresses 4 and 6 respectively, write a program in Z80 code (including flowchart) to make the display count up slowly in binary.

Flowchart: see *Figure 15*.

189

Program code:

Machine code		Assembly language
	ORG 0C90H	
0C90 3E 0F	LD A,0FH	
0C92 D3 06	OUT (6),A	; Port A byte output mode
0C94 21 00 01	LOOP1 LD HL,0100H	; set up delay time
0C97 1C	LOOP2 INC E	; time delay
0C98 20 FD	JR NZ LOOP2	
0C9A 2B	DEC HL	
0C9B 7D	LD A,L	
0C9C B4	OR H	; check for HL=0 (set Z flag)
0C9D 20 F8	JR NZ, LOOP2	
0C9F 78	LD A,B	; transfer binary counter to A
0CA0 D3 04	OUT (4),A	; update LED display
0CA2 04	INC B	; increment counter by 1
0CA3 18 EF	JR LOOP1	

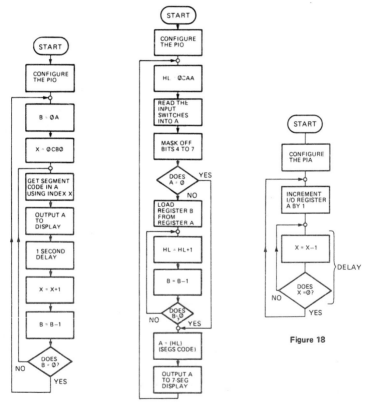

Figure 16

Figure 17

Figure 18

190

Problem 9 A Z80 based microcomputer with a Z80 PIO is connected to four SPST switches and a 7-segment LED display using the interface circuits shown in *Figures 2 and 10*. Assuming Port A data and mode control registers use I/O addresses 4 and 6 respectively and Port B data and mode control registers use I/O addresses 5 and 7 respectively, write a program in Z80 code (including flowchart), to display the binary value selected by the switches as a hexadecimal character on the 7-segment display.

Flowchart: see *Figure 16*.

Program code:

Machine code		Assembly language	
	ORG 0C90H		
0C90 3E 0F		LD A,0FH	
0C92 D3 06		OUT (6),A	; Port A byte output mode
0C94 3E 4F		LD A,4FH	
0C96 D3 07		OUT (7),A	; Port B byte input mode
0C98 21 AA 0C	READ	LD HL,SEGS	; HL points to segments table
0C9B DB 05		IN A,(5)	; read the switches
0C9D E6 0F		AND 0FH	; mask off bits 4 to 7
0C9F 28 04		JR Z,DISP	; no conversion if zero
0CA1 47		LD B,A	; swap registers
0CA2 23	CONV	INC HL	; advance HL through segs table
0CA3 10 FD		DJNZ CONV	; if B=0, HL points to segments
			; code
0CA5 7E	DISP	LD A,(HL)	; get segments code
0CA6 D3 04		OUT (4),A	; and send to display
0CA8 18 EE		JR READ	; keep repeating
0CAA 3F	SEGS	DEFB 3FH,06H,5BH; 0,1,2	
0CAB 06			
0CAC 5B			
0CAD 4F		DEFB 4FH,66H,6DH; 3,4,5	
0CAE 66			
0CAF 6D			
0CB0 7D		DEFB 7DH,07H,7FH; 6,7,8	
0CB1 07			
0CB2 7F			
0CB3 67		DEFB 67H,77H,7CH;9,A,B	
0CB4 77			
0CB5 7C			
0CB6 39		DEFB 39H,5EH,79H; C,D,E	
0CB7 5E			
0CB8 79			
0CB9 71		DEFB 71H ; F	

Flowchart: see *Figure 17*.

Program code:

Machine code			Assembly language
		ORG ØC9ØH	
ØC90 3E ØF		LD A,ØFH	
ØC92 D3 Ø6		OUT (6),A	; Port A byte output mode
ØC94 Ø6 ØA	RESET	LD B,ØAH	; load B as a decade counter
ØC96 DD 21 BØ ØC		LD IX,SEGS	; IX points to segments table
ØC9A DD 7E ØØ	COUNT	LD A,(IX+Ø)	; get a segments code
ØC9D D3 Ø4		OUT (4),A	; display on LED
ØC9F 21 ØØ Ø2		LD HL,Ø2ØØH	; set up delay time
ØCA2 1C	DELAY	INC E	; 1 second delay
ØCA3 2Ø FD		JR NZ,DELAY	
ØCA5 2B		DEC HL	
ØCA6 7D		LD A,L	
ØCA7 B4		OR H	; check for HL=Ø (set Z flag)
ØCA8 2Ø F8		JR NZ,DELAY	
ØCAA DD 23		INC IX	; advance IX through segs.
			; table
ØCAC 1Ø EC		DJNZ COUNT	; display next digit
ØCAE 18 E4		JR RESET	; start again if 9 is exceeded
ØCBØ 3F	SEGS	DEFB 3FH,Ø6H,5BH; Ø,1,2	
ØCB1 Ø6			
ØCB2 5B			
ØCB3 4F		DEFB 4FH,66H,6DH; 3,4,5	
ØCB4 66			
ØCB5 6D			
ØCB6 7D		DEFB 7DH,Ø7H,7FH; 6,7,8	
ØCB7 Ø7			
ØCB8 7F			
ØCB9 67		DEFB 67H ; 9	

Flowchart: see *Figure 18*.

Program code:

Machine code		Assembly language
	DRA EQU $8004	
	CRA EQU $8005	
	ORG $0020	
0020 86 FF	LDA A #$FF	
0022 B7 80 04	STA A DRA	; configure data direction reg. A
0025 86 04	LDA A #$04	
0027 B7 80 05	STA A CRA	; select I/O register A
002A 7C 80 04 COUNT	INC DRA	; use I/O reg. as a binary counter
002D 09 DELAY	DEX	
002E 26 FD	BNE DELAY	
0030 20 F8	BRA COUNT	

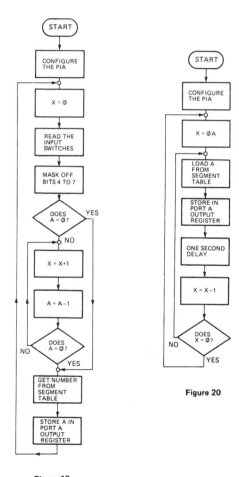

Figure 19

Figure 20

193

Flowchart: see *Figure 19*.

Program code:

Machine code		DRA EQU $8004	Assembly language
		CRA EQU $8005	
		DRB EQU $8006	
		CRB EQU $8007	
		ORG $0020	
0020 86 FF		LDA A #$FF	
0022 B7 80 04		STA A DRA	; Port A all outputs
0025 4F		CLR A	
0026 B7 80 06		STA A DRB	; Port B all inputs
0029 86 04		LDA A #$04	
002B B7 80 05		STA A CRA	; select I/O register A
002E B7 80 07		STA A CRB	; select I/O register B
0031 CE 00 00	READ	LDX #$0000	
0034 B6 80 06		LDA B DRB	; read the switches
0037 84 0F		AND #$0F	; mask off bits 4 to 7
0039 27 04		BEQ DISP	; no conversion if zero
003B 08	CONV	INX	; advance X through segs table
003C 4A		DEC A	
003D 26 FC		BNE CONV	; X points to segs code if A=0
003F A6 46	DISP	LDA A SEGS,X	; get segs code
0041 B7 80 04		STA A DRA	; and send to display
0044 20 EB		BRA READ	; keep repeating
0046 3F	SEGS	FCB $3F,$06,$5B	; 0, 1, 2
0047 06			
0048 5B			
0049 4F		FCB $4F,$66,$6D	; 3, 4, 5
004A 66			
004B 6D			
004C 7D		FCB $7D,$07,$7F	; 6, 7, 8
004D 07			
004E 7F			
004F 67		FCB $67,$77,$7C	; 9, A, B
0050 77			
0051 7C			
0052 39		FCB $39,$5E,$79	; C, D, E
0053 5E			
0054 79			
0055 71		FCB $71	; F

194

Problem 13 A 6800 based microcomputer with a 6820 PIA is connected to a single 7-segment display using the interface circuit shown in *Figure 10* of this chapter. Assuming the PIA is located in the address range $8004 to $8007, write a program in 6800 code (including flowchart), to display a count of 0 to 9 repetitively at one second intervals.

Flowchart: see *Figure 20*.

Program code

Machine code				Assembly language
	DELAY1	EQU	$0080	
	DELAY2	EQU	$0081	
	DRA	EQU	$8004	
	CRA	EQU	$8005	
		ORG	$0020	
0020 86 FF		LDA A	#$FF	
0022 B7 80 04		STA A	DRA	; Port A all outputs
0025 86 04		LDA A	#$04	
0027 B7 80 05		STA A	CRA	; select I/O register A
002A CE 00 0A	RESET	LDX	#$0A	; point X to end of table
002D A6 45	COUNT	LDA A	TABLE,X	; get a segments code
002F B7 80 04		STA A	DRA	; light up display
0032 C6 01		LDA B	#$01	; delay constant
0034 7A 00 80	LOOP	DEC	DELAY1	; 1 second delay
0037 26 FB		BNE	LOOP	
0039 7A 00 81		DEC	DELAY2	
003C 26 F6		BNE	LOOP	
003E 5A		DEC B		
003F 26 F3		BNE	LOOP	
0040 09		DEX		; move X to next code position
0041 26 E9		BNE	COUNT	; carry on counting if not past 9
0043 20 E4		BRA	RESET	; otherwise start from 0 again
0045 67	TABLE	FCB	$67,$7F,$07	; 9, 8, 7
0046 7F				
0047 07				
0048 7D		FCB	$7D,$6D,$66	; 6, 5, 4
0049 6D				
004A 66				
004B 4F		FCB	$4F,$5B,$06	; 3, 2, 1
004C 5B				
004D 06				
004E 3F		FCB	$3F	; 0

C   FURTHER PROBLEMS ON INTERFACING

1   External devices, attached to a microcomputer to enable communication with the outside world to take place, are called . . . .

2   A circuit which allows two otherwise incompatible systems to be interconnected is called an . . . .

3   Voltage or current amplifiers, connected between a microcomputer output and its external devices, are known as . . . .

4   Signal lines, connected between a microcomputer and its external devices, whose function is to control the timing of data transfers are called . . . .

5   In order to connect a 7-segment LED display to the output of a microcomputer, ... and . . . . are necessary.

6   A special purpose, programmable integrated circuit, used in a microcomputer to facilitate the connection of external devices, is known as a . . . .

(b) CONVENTIONAL PROBLEMS

1   A 6502 based microcomputer with a 6530 PIA is connected to eight LEDs using an interface circuit of the type shown in *Figure 2* of this chapter. Write a program in 6502 code (including flowchart), to make the LEDs form a 'running light' display, i.e. each LED switching on and off in sequence from one end of the display to the other.

2   A 6502 based microcomputer with a 6530 PIA is connected to eight LEDs and four SPST switches using an interface circuit of the type shown in *Figure 2* of this chapter. Write a program in 6502 code (including flowchart), to form an up/down binary counter, such that the position of one of the SPST switches determines whether the LED display indicates an up count or a down count. A counting rate of approximately one count per second should be used.

3   A 6502 based microcomputer with a 6530 PIA is connected to four SPST switches and a single 7-segment LED display using interface circuits of the type shown in *Figures 2 and 10*. Write a program in 6502 code (including flowchart), to form an up/down decade counter, such that the position of one of the SPST switches determines whether the display indicates an up count or a down count. A counting rate of one digit per second should be used.

4 A 6502 based microcomputer with a 6530 PIA is connected to four SPST switches and a single 7-segment LED display using interface circuits of the type shown in *Figures 2 and 10*. Write a program in 6502 code (including flowchart), such that each closure of one of the SPST switches causes the displayed digit to be incremented by 1, and each closure of a second SPST switch causes the displayed digit to be decremented by 1.

5 A 6502 based microcomputer with a 6530 PIA is connected to six LEDs using an interface circuit of the type shown in *Figure 2*. Three different coloured LEDs are used to represent two sets of traffic lights at a road junction. Write a program in 6502 code (including flowchart), to operate the lights to control traffic flow in two directions. (Hint: the use of a look-up table may provide the simplest solution).

6 A Z80 microcomputer with a Z80 PIO is connected to eight LEDs using an interface circuit of the type shown in *Figure 2*. Write a program in Z80 code (including flowchart), to make the LEDs form a 'running light' display, i.e. each LED switching on and off in sequence from one end of the display to the other.

7 A Z80 based microcomputer with a Z80 PIO is connected to eight LEDs and four SPST switches using an interface circuit of the type shown in *Figure 2*. Write a program in Z80 code (including flowchart), to form an up/down binary counter, such that the position of one of the SPST switches determines whether the LED display indicates an up count or a down count. A counting rate of approximately one count per second should be used.

8 A Z80 based microcomputer with a Z80 PIO is connected to four SPST switches and a single segment LED display using an interface circuit of the type shown in *Figures 2 and 10*. Write a program in Z80 code (including flowchart), to form an up/down decade counter, such that the position of one of the SPST switches determines whether the display indicates an up count or a down count. A counting rate of approximately one digit per second should be used.

9 A Z80 based microcomputer with a Z80 PIO is connected to four SPST switches and a single 7-segment LED display using interface circuits of the type shown in *Figures 2 and 10*. Write a program in Z80 code (including flowchart), such that each closure of one of the SPST switches causes the displayed digit to be incremented by 1, and each closure of a second SPST switch causes the displayed digit to be decremented by 1.

10 A Z80 based microcomputer with a Z80 PIO is connected to six LEDs using an interface circuit of the type shown in *Figure 2*. Three different coloured LEDs are used to represent two sets of traffic lights at a road junction. Write a program in Z80 code (including flowchart), to operate the lights to control traffic flow in two different directions. (Hint: the use of a look-up table may provide the simplest solution).

11 A 6800 based microcomputer with a 6820 PIA is connected to eight LEDs using an interface circuit of the type shown in *Figure 2*. Write a program in 6800 code (including flowchart), to make the LEDs form a 'running light' display, i.e. each LED switching on and off in sequence from one end of the display to the other.

12 A 6800 based microcomputer with a 6820 PIA is connected to eight LEDs and four SPST switches using an interface circuit of the type shown in *Figure 2*. Write

a program in 6800 code (including flowchart), to form an up/down binary counter, such that the position of one of the SPST switches determines whether the LED display indicates an up count or a down count. A counting rate of approximately one count per second should be used.

13 A 6800 based microcomputer with a 6820 PIA is connected to four SPST switches and a single segment LED display using an interface circuit of the type shown in *Figures 2 and 10*. Write a program in 6800 code (including flowchart), to form an up/down decade counter, such that the position of one of the SPST switches determines whether the display indicates an up count or a down count. A counting rate of approximately one digit per second should be used.

14 A 6800 based microcomputer with a 6820 PIA is connected to four SPST switches and a single 7-segment LED display using interface circuits of the type shown in *Figures 2 and 10*. Write a program (including flowchart), such that each closure of one of the SPST switches causes the displayed digit to be incremented by 1, and each closure of a second SPST switch causes the displayed digit to be decremented by 1.

15 A 6800 based microcomputer with a 6820 PIA is connected to six LEDs using an interface circuit of the type shown in *Figure 2*. Three different coloured LEDs are used to represent two sets of traffic lights at a road junction. Write a program in 6800 code (including flowchart), to operate the lights to control traffic flow in two different directions. (*Hint: the use of a look-up table may provide the simplest solution.*)

7 Subroutines and the stack

A MAIN POINTS CONCERNED WITH SUBROUTINES AND THE STACK

1 Sequences of instructions which perform general purpose functions such as mathematical processes, code conversions or input/output data transfers are known as **'routines'**. Such routines may need to be used several times at different places throughout a microcomputer program. It is therefore advantageous if frequently used routines can be written in such a way that they may be called from any point in a program, as many times as necessary, without having to repeat their instructions. Routines which are written so that they may be used in this manner are known as **'subroutines'**, and the basic mechanism of a subroutine is shown in *Figure 1*.

2 Two instructions are provided in most microprocessor instruction sets to enable subroutines to be implemented, and these are:
   (i)   **'JSR' (jump to subroutine)** or **'CALL'** which are used in programs to enable a subroutine to be called. The effect of executing these instructions is to allow the program counter to locate a subroutine by loading it with the starting address of the subroutine being called; and
   (ii) **'RTS' (return from subroutine)** or **'RET' (return)** which are used as the last instruction in a subroutine, and which allow a return to the calling program at the correct point by loading the program counter with the address of the next logical instruction (i.e. the instruction in the calling program which follows the JSR or CALL instruction).
In some microprocessors e.g. Z80, both CALL and RET instructions may be made conditional.

3 As a result of using subroutines in a program, the following characteristics are achieved:
   (i)   **shorter object code** (machine code), since it is not necessary to repeatedly write out the code used for a particular routine,
   (ii) **improved program structure**, since programs may be constructed by selecting from a library of fully debugged subroutines. This modular approach provides greater flexibility and allows changes to be readily implemented,
   (iii) **improved program readability**, since the main program consists of a sequence of calls to readily identifiable functions,
   (iv) **increased running time**, since additional instructions are required to call and return from each subroutine, and these instructions have relatively long execution times (see *Appendix B*, page 250), and

MAIN PROGRAM

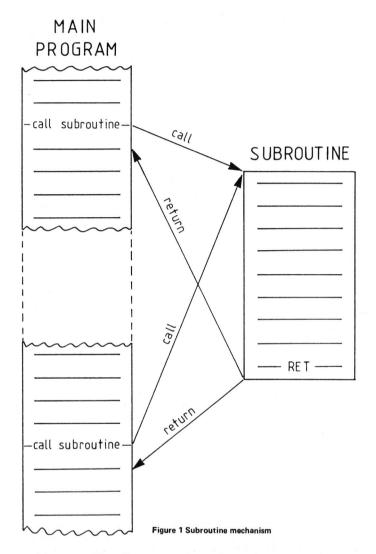

Figure 1 Subroutine mechanism

(v) **corruption of important registers**, since registers which are in use in a main program may be required for use in a subroutine. Upon returning from a subroutine, therefore, such registers may contains values which are determined by the subroutine and their previous contents are lost.

4 Since a subroutine may be called from many different points in a program, it is necessary to save a **'return address'**, so that control may be passed back to the program at the correct point upon completion of the subroutine. Return

addresses are stored in a **'last-in-first-out'** (LIFO) register which is known as a **'stack'**. There are two ways in which a stack may be implemented, and these are:

(i) an internal or **hardware stack**, in which a block of registers are included within a microprocessor for the purpose of saving return addresses, and

(ii) an external or **software stack**, in which a section of RAM is used for saving return addresses and a **'stack pointer'** register in the microprocessor is used to point to the end of the stack (see *Worked Problems 1 to 4*).

Since a hardware stack is limited in size, most microprocessors use a software stack. Return addresses are automatically saved on the stack when a subroutine is called, and retrieved from the stack when a return instruction is encountered at the end of a subroutine.

5   It is possible to call a subroutine from within another subroutine. Such an arrangement is known as a **'nested subroutine'**, and the use of nested subroutines is shown in *Figure 2*. For each additional level of subroutine nesting

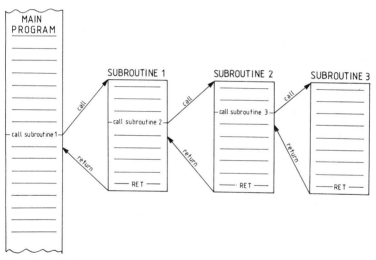

Figure 2 Nested subroutines

used, an additional return address must be saved on the stack, thus causing the stack to be extended. Therefore, the degree of nesting possible in a particular microcomputer is determined by the size of the stack available to the user.

6   In addition to the automatic use of the stack when subroutines are used, a stack may be used manually by a programmer for the temporary storage of data. Two instructions are available in most microprocessor instruction sets for this purpose, and these are:

(i) **PUSH** which causes data in specified registers to be saved in the next free location on the stack, and

(ii)   **PULL** or **POP**, which causes data to be retrieved from the last used location on the stack and transferred to a specified register.

7   It is often important that data contained in the various registers of a microprocessor are not disturbed by the action of calling and executing a subroutine. Certain registers may be required for use in a subroutine, and it is therefore important that the contents of these registers are saved before entering the subroutine, and are restored upon returning to the main program. Data in these registers may be saved by one of the following methods:

(i)   temporarily transfer data to **other registers** which are not in use in the main program or in the subroutine,
(ii)   save the data in **specified memory locations,**
(iii) transfer the data to a **alternate register set** if available (e.g. Z80), and
(iv)   save the data on the **stack.**

Temporarily saving the contents of registers in this manner is known as **'saving the status'** of the program.

8   In most cases where subroutines are used, the data processed by the subroutine is different each time that the subroutine is called. Therefore it is necessary for data to be passed from the main program to the appropriate subroutine prior to calling it, and for results to be passed from the subroutine back to the main program before leaving the subroutine. This is a technique known as **'parameter passing',** and this may be accomplished in one of the following ways:

(i)   by passing data via **specified register**,
(ii)   by passing data via **specified memory locations**, and
(iii) by passing data via the **stack**.

B WORKED PROBLEMS ON SUBROUTINES AND THE STACK

Problem 1   Describe the sequence of events which occur when a subroutine is called by a main program.

A typical sequence of events which occur when a subroutine is called is as follows:

(i)   **push PCH onto the stack,**
(ii)   **decrement the stack pointer,**
(iii) **push PCL onto the stack,**
(iv)   **decrement the stack pointer,**
(v)   **transfer the subroutine starting address to PCH and PCL,** and
(vi)   **start execution of the subroutine**.

This sequence of events is illustrated in *Figure 3*.

Problem 2   Describe the sequence of events which occur at the end of a subroutine when a return to the main program is initiated by an RTS or RET instruction.

A typical sequence of events which occur during a return to the main program from a subroutine is as follows:
(i)   **increment the stack pointer;**
(ii)   **pull PCL from the stack;**

(iii)   **increment the stack pointer;**
(iv)   **pull PCH from the stack;** and
(v)   **continue execution of the main program.**

This sequence of events is illustrated in *Figure 4*.

Problem 3   Explain the importance of initializing the stack pointer of a microprocessor when writing programs which make use of subroutines, and show how this may be carried out when using the following microprocessors: (a) **6502**; (b) **Z80**; (c) **6800**.

The stack is an area of RAM where data is stored on a **'last-in-first-out'** basis with the aid of a **stack pointer** register. Each time that data is stored on the stack, the stack is extended downwards in memory from a starting address determined by the initial contents of the stack pointer. Unless the stack pointer is loaded with a predetermined value at the start of a user's program, the stack may be located:

(i)   in a position where there is no RAM (e.g. where there is ROM), in which case data will not be saved and correct returns from subroutines will not be achieved, or

(ii)   close to, or within, the area of RAM used to store user's programs, in which case the use of the stack may corrupt the user's program by overwriting it, and cause the program to crash.

The stack pointers of the 6502, Z80 and 6800 microprocessors may be initialized in the following manner:

(a)   The upper eight bits of the stack pointer of a **6502 stack pointer** are always preset to 01 by the microprocessor itself, and only the lower eight bits may be determined by the user. This means that a 6502 stack must reside in page 1 (addresses 0100_{16} to $01FF_{16}$), and provided that user programs are kept out of this area of memory, crashes due to the stack overwriting the user's program cannot occur. This arrangement limits the size of the stack, and nesting of subroutines to a greater depth than 128 is not possible, although this is more than adequate for most programs. The usual location for the top of the stack is address $01FF_{16}$, and the stack pointer may be set to this address in the following manner:

```
A2 FF    LDX #$FF    ;load X with stack top address
9A       TXS         ;and transfer to stack pointer
```

Note the use of index register X, since the 6502 has no direct instruction to load its stack pointer.

(b)   The **Z80 stack pointer** may be loaded directly with any 16 bit address, and a convenient location for the stack in a Z80 system is at the top end of the user's RAM. Since the stack pointer in a Z80 microprocessor always points to the last used location of the stack (as opposed to the next free location, as is the case with the 6502 and 6800 microprocessors), a suitable address with which to initialize the stack pointer is one higher than the highest RAM address in the system. For example, if the highest address in RAM on a system is $0FFF_{16}$, the stack pointer may be initialized as follows:

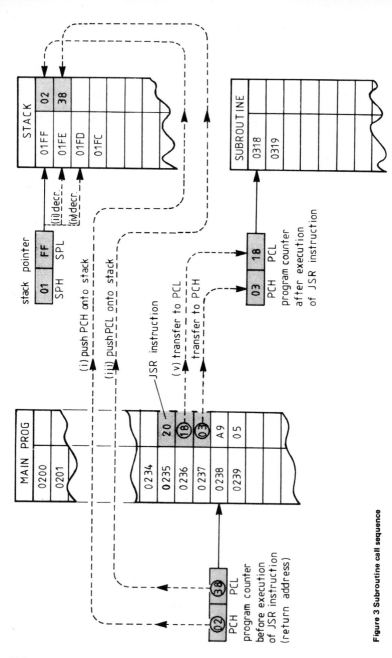

Figure 3 Subroutine call sequence

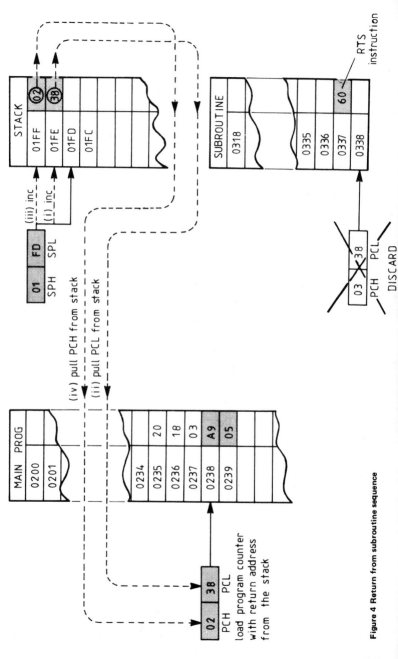

Figure 4 Return from subroutine sequence

```
31 ØØ 1Ø     LD SP, 1ØØØH  ;point stack pointer to top of RAM
                           ;plus 1
```

(c)   Like the Z80 stack pointer, the **6800 stack pointer** may be loaded directly
       with any 16 bit address, but unlike the Z80, the 6800 stack pointer always
       points to the next free location on the stack. Any convenient area of
       RAM may be used for the stack which is clear of the user's program area.
       For example, if a section of RAM is available from $AØØØ_{16}$ to $AØ7F_{16}$,
       the stack pointer may be initialized as follows:

```
8E AØ 78     LDS #$AØ78
```

Note that in the absence of these instructions, most operating (monitor) systems
load the stack pointer with a default value after a system reset.

Problem 4   A section of a program which uses nested subroutines is
illustrated in *Figure 5*.
    Construct suitable tables to show the behaviour of the stack during this
part of the program, assuming the stack pointer is initialized at address
$Ø1FF_{16}$ and each subroutine call instruction is three bytes in length.

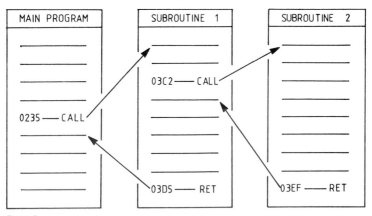

Figure 5

The operation of the stack during execution of this program is shown in *Figure 6*.

Problem 5   Show how the important registers in a **6502** microprocessor may
be:
(a)   saved **prior to calling** a subroutine, and
(b)   restored **after leaving** a subroutine.

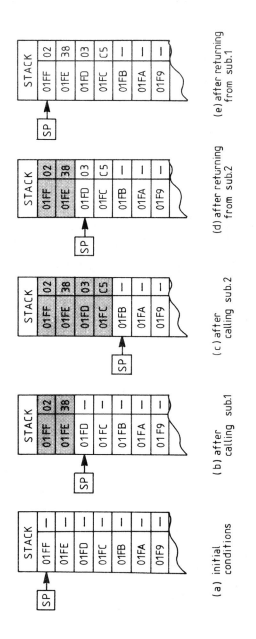

(a) initial conditions

(b) after calling sub.1

(c) after calling sub.2

(d) after returning from sub.2

(e) after returning from sub.1

Note: the stack pointer (SP) of a Z80 MPU points to the last used location on the stack rather than the next free location, as show above for the 6502 & 6800 MPUs.

Figure 6

207

(a)

Ø8	PHP	;save the status register on the stack
48	PHA	;save the accumulator on the stack
8A	TXA	
48	PHA	;save index register X on the stack
98	TYA	
48	PHA	;save index register Y on the stack

(b)

68	PLA	
A8	TAY	;restore index register Y from the stack
68	PLA	
AA	TAX	;restore index register X from the stack
68	PLA	;restore the accumulator from the stack
28	PLP	;restore the status register from the stack

Problem 6 A **6502** machine code subroutine is required to provide a **time delay** of between 1 and 10 seconds, depending upon the value stored in **index register X** prior to calling the subroutine.

Write a suitable subroutine, starting at address $\emptyset 3\emptyset\emptyset_{16}$ to enable the delay times required to be covered. The value in the accumulator prior to calling the subroutine should be preserved.

```
                    ;Time delay subroutine (DELAY)
                    ;Entry: X contains the delay parameter
                    ;Exit: X=ØØ
                    ;A and Y are unchanged
                    ;
                    COUNT1    =$ØØ8Ø
                    COUNT2    =$ØØ81
                         *    =$Ø3ØØ
                    ;
Ø3ØØ   48    DELAY   PHA              ;save accumulator
Ø3Ø1   A9 2Ø        LDA #$2Ø         ;
Ø3Ø3   85 8Ø        STA COUNT1       ;preset counter 1
Ø3Ø5   85 81        STA COUNT2       ;preset counter 2
Ø3Ø7   C6 81 LOOP   DEC COUNT2
Ø3Ø9   DØ FC        BNE LOOP         ;keep looping if not zero
Ø3ØB   C6 8Ø        DEC COUNT1
Ø3ØD   DØ F8        BNE LOOP         ;keep looping if not zero
Ø3ØF   CA           DEX              ;decrement delay parameter
Ø31Ø   DØ EF        BNE LOOP         ;keep looping if not zero
Ø312   68           PLA              ;restore A from stack
Ø313   6Ø           RTS              ;return to calling program
```

Table 1 shows the values required in index register X, prior to calling this subroutine, to obtain time delays between 1 and 10 seconds. These values assume that the program is used in a 6502 based microcomputer with a 1 MHz clock.

Time delay (seconds)	Value in X
1	10
2	1F
3	2F
4	3F
5	4E
6	5E
7	6D
8	7D
9	8D
10	9D

Table 1

Problem 7 A **6502** machine code subroutine is required to convert the **BCD value** stored in b0–b3 of the accumulator into its equivalent **7-segment code**.
   Write a suitable subroutine, starting at address 0314_{16}, to perform the necessary code conversion and return with the segments code in the accumulator.

```
                    ;BCD to 7-segment conversion routine (SEG7)
                    ;Entry: BCD data in A
                    ;Exit: segments code in A
                    ;X is changed, Y is unchanged
                    ;
                          *  =$0314
                    ;
0314   AA      SEG7    TAX              ;use value in A as index
0315   BD 19 03        LDA TABLE,X      ;get segments code
0318   60              RTS              ;return to calling program
0319   3F      TABLE   .BYTE $3F,$06,$5B; 0, 1, 2
031A   06
031B   5B
031C   4F              .BYTE $4F,$66,$6D; 3, 4, 5
031D   66
031E   6D
031F   7D              .BYTE $7D,$07,$7F; 6, 7, 8
0320   07
0321   7F
0322   67              .BYTE $67      ; 9
```

Problem 8 A **6502** machine code subroutine is required to convert a **packed BCD** value in index register X into its ASCII equivalents and store these in memory locations 0082_{16} and 0083_{16}.

Write a suitable subroutine, starting at address 0323_{16}, to perform the necessary code conversion. The value in the accumulator should remain unchanged as a result of calling this subroutine.

```
                ;BCD to ASCII subroutine (BCDASC)
                ;Entry: BCD data in X
                ;Exit: ASCII codes in $0082 and $0083
                ;A,X and Y are unchanged
                ;
                ASCLO    =$0082
                ASCHI    =$0083
                    *    =$0323
                ;
0323  48        BCDASC  PHA             ;save A on the stack
0324  8A                TXA             ;get BCD into A
0325  29 0F             AND#$0F         ;mask off high nibble
0327  09 30             ORA #$30        ;convert to ASCII
0329  85 82             STA ASCLO       ;and save
032B  8A                TXA             ;get BCD into A again
032C  4A                LSR             ;shift low nibble
032D  4A                LSR             ;4 places to the left
032E  4A                LSR             ;into high nibble position
032F  4A                LSR             ;
0330  09 30             ORA #$30        ;convert to ASCII
0332  85 83             STA ASCHI       ;and save
0334  68                PLA             ;restore A
0335  60                RTS             ;return from subroutine
```

Problem 9 A **6502** machine code subroutine is required to convert **two**
ASCII codes in memory locations 0082_{16} and 0083_{16} into **packed BCD** form.
   Write a suitable subroutine, starting at address 0336_{16}, to perform the
necessary code conversion and store the packed BCD equivalent in index
register X. The value in the accumulator should remain unchanged as a
result of calling this subroutine.

```
                    ;ASCII to packed BCD subroutine (ASCBCD)
                    ;Entry: ASCII data in $0082 and $0083
                    ;Exit: packed BCD in X
                    ;A and Y are unchanged
                    ;
                    ASCLO    =$0082
                    ASCHI    =$0083
                         *   =$0336
                    ;
0336   48    ASCBCD  PHA              ;save A on the stack
0337   A5 82         LDA $0082        ;get low nibble ASCII code
0339   29 0F         AND #$0F         ;convert ASCII to BCD
033B   85 82         STA $0082        ;save temporarily
033D   A5 83         LDA $0083        ;get high nibble ASCII code
033F   0A            ASL A            ;convert ASCII to BCD
0340   0A            ASL A            ;and shift into high
0341   0A            ASL A            ;nibble position
0342   0A            ASL A            ;
0343   05 82         ORA $0082        ;combine high an low nibbles
0345   AA            TAX              ;and pack into X
0346   68            PLA              ;restore A
0347   60            RTS              ;return from subroutine
```

Problem 10 A **6502** machine code subroutine is required to **multiply** together two numbers stored in the accumulator and index register X, and store the product in memory locations 0084_{16} and 0085_{16}.
   Write a suitable subroutine, starting at address 0348_{16}.

```
                ;Multiplication subroutine (MULT)
                ;Entry: multiplier in X, multiplicand in A
                ;Exit: product high in A and in $0084
                ;      product low in $0085
                ;Y is unchanged
                ;
                PRODHI  =$0084
                PRODLO  =$0085
                     *  =$0348
                ;
0348  85 84  MULT    STA PRODHI   ;transfer multiplicand to memory
034A  86 85          STX PRODLO   ;transfer multiplier to memory
034C  A9 00          LDA #$00     ;clear the accumulator
034E  A2 08          LDX #$08     ;use X as loop counter
0350  6A     SHIFT   ROR A        ;shift product/multiplier right
0351  66 85          ROR PRODLO   ;one place
0353  90 03          BCC NOADD    ;skip add if multiplier bit is 0
0355  18             CLC
0356  65 84          ADC PRODHI   ;add multiplicand to product
0358  CA     NOADD   DEX          ;decrement loop counter
0359  D0 F5          BNE SHIFT    ;repeat if not done 8 times
035B  4A             LSR A        ;final shift
035C  66 85          ROR PRODLO   ;of product
035E  85 84          STA PRODHI
0360  60             RTS          ;return to calling program
```

Problem 11 A **6502** machine code routine is required to **divide** a number in the accumulator by the number stored in index register X.

Write a suitable subroutine, starting at address $\emptyset361_{16}$ to divide the two numbers and save the quotient and remainder in memory locations $\emptyset\emptyset86_{16}$ and $\emptyset\emptyset87_{16}$.

```
                    ;Division subroutine (DIVIDE)
                    ;Entry: divisor in X, dividend in A
                    ;Exit: quotient in $∅∅86
                    ;      remainder in A and in $∅∅87
                    ;Y is unchanged
                    ;
                    QUOTNT =$∅∅86
                    REMNDR =$∅∅87
                         * =$∅361
                    ;
∅361   85 86   DIVIDE   STA QUOTNT   ;transfer dividend to memory
∅363   86 87            STX REMNDR   ;transfer divisor to memory
∅365   A9 ∅∅            LDA #$∅∅     ;clear the accumulator
∅367   A2 ∅8            LDX #$∅8     ;use X as loop counter
∅369   26 86   SHIFT    ROL QUOTNT   ;shift dividend/quotient
∅36B   2A               ROL A        ;one place to the left
∅36C   38               SEC          ;
∅36D   E5 87            SBC REMNDR   ;subtract the divisor
∅36F   B∅ ∅3            BCS NOADD    ;do not restore if C=1
∅371   65 87            ADC REMNDR   ;divisor too big, so restore
∅373   18               CLC          ;
∅374   CA      NOADD    DEX          ;decrement loop counter
∅375   D∅ F2            BNE SHIFT    ;repeat if not done 8 times
∅377   26 86            ROL QUOTNT   ;final shift of quotient
∅379   85 87            STA REMNDR   ;transfer remainder to memory
∅37B   6∅               RTS          ;return to calling program
```

213

```
                    ;Binary to BCD conversion subroutine (BIDEC)
                    ;Entry: binary number in A
                    ;Exit: BCD equivalent in $ØØ89 and $ØØ8A
                    ;
                    TEMP      =$ØØ88
                    BCDLO     =$ØØ89
                    BCDHI     =$ØØ8A
                          *   =$Ø37C
                    ;
Ø37C   F8    BIDEC   SED             ;use decimal mode
Ø37D   85 88         STA TEMP        ;save A temporarily
Ø37F   8A            TXA
Ø38Ø   48            PHA             ;save X on the stack
Ø381   A2 Ø8         LDX #$Ø8        ;use X as a loop counter
Ø383   A9 ØØ         LDA #$ØØ        ;clear A
Ø385   85 89         STA BCDLO       ;clear BCDLO
Ø387   Ø6 88 SHIFT   ASL TEMP        ;move each bit into carry flag
Ø389   A5 89         LDA BCDLO       ;
Ø38B   65 89         ADC BCDLO       ;double and add carry
Ø38D   85 89         STA BCDLO       ;and put back
Ø38F   26 8A         ROL BCDHI       ;shift any carry into BCDHI
Ø391   CA            DEX             ;decrement loop counter
Ø392   DØ F3         BNE SHIFT       ;and repeat if not done 8 times
Ø394   68            PLA             ;
Ø395   AA            TAX             ;restore X
Ø396   D8            CLD             ;restore binary mode
Ø397   6Ø            RTS             ;return to calling program
```

Problem 13 A **6502** machine code subroutine is required to convert a **decimal (BCD)** number in the accumulator into its **binary** equivalent and store this in memory location $\emptyset\emptyset8B_{16}$.

Write a suitable subroutine, starting at address $\emptyset398_{16}$, to perform this function.

```
                    ;BCD to binary subroutine (DECBI)
                    ;Entry: BCD number in A
                    ;Exit: binary number in $∅∅8B
                    ;X and Y are unchanged
                    ;
                    MASK    =$∅∅88
                    BINARY  =$∅∅8B
                         *  =$∅398
                    ;
∅398   F8    DECBI    SED             ;use decimal mode
∅399   85 8B          STA BINARY      ;use as temp store for A
∅39B   8A             TXA
∅39C   48             PHA             ;save X on the stack
∅39D   A9 ∅8          LDA #$∅8        ;load A with a mask
∅39F   85 88          STA MASK        ;and store in memory
∅3A1   A5 8B          LDA BINARY      ;restore A
∅3A3   A2 ∅8          LDX #$∅8        ;use X as a loop counter
∅3A5   4A    SHIFT    LSR A           ;divide A by 2
∅3A6   66 8B          ROR BINARY      ;shift remainder into BINARY
∅3A8   24 88          BIT MASK        ;see if A is illegal BCD
∅3AA   F∅ ∅3          BEQ HCAR        ;skip correction if not
∅3AC   38             SEC
∅3AD   E9 ∅3          SBC #$∅3        ;otherwise apply correction
∅3AF   CA    HCAR     DEX             ;decrement loop counter
∅3B∅   D∅ F3          BNE SHIFT       ;repeat if not done 8 times
∅3B2   68             PLA
∅3B3   AA             TAX             ;restore X
∅3B4   D8             CLD             ;restore binary mode
∅3B5   6∅             RTS             ;return to calling program
```

Problem 14 Refer to the circuit diagram of a **6-digit multiplexed LED** display shown in *Figure 7*.

Write a **6502** machine code subroutine starting at address $\emptyset3B6_{16}$ to display the BCD contents of a display buffer, located in memory at $\emptyset\emptyset8D_{16}$, $\emptyset\emptyset8E_{16}$ and $\emptyset\emptyset8F_{16}$. All register contents should be preserved.

```
                    ;Multiplexed display subroutine (MUXDSP)
                    ;Entry: BCD codes in $ØØ8D, $ØØ8E & $ØØ8F
                    ;A, X and Y are preserved
                    ;This subroutine calls DSPLY subroutine at $Ø3D1
                    ;
                    DISPNO  =$ØØ8C
                    DISBUF  =$ØØ8D
                       *    =$Ø3B6
                    ;
Ø3B6  48      MUXDSP PHA              ;save A on the stack
Ø3B7  98             TYA
Ø3B8  48             PHA              ;save Y on the stack
Ø3B9  8A             TXA
Ø3BA  48             PHA              ;save X on the stack
Ø3BB  A9 ØØ          LDA #$ØØ
Ø3BD  85 8C          STA DISPNO       ;initialise display counter
Ø3BF  A2 ØØ          LDX #$ØØ         ;clear buffer index
Ø3C1  B5 8D   DISPCD LDA DISBUF,X     ;get code from buffer
Ø3C3  2Ø D1 Ø3       JSR DSPLY        ;light up one pair of digits
Ø3C6  E8             INX              ;point to next position in buffer
Ø3C7  EØ Ø3          CPX #$Ø3         ;check for buffer end
Ø3C9  DØ F6          BNE DISPCD       ;carry on if not at buffer end
Ø3CB  68             PLA
Ø3CC  AA             TAX              ;restore X
Ø3CD  68             PLA
Ø3CE  A8             TAY              ;restore Y
Ø3CF  68             PLA              ;restore A
Ø3DØ  6Ø             RTS              ;return to calling program
```

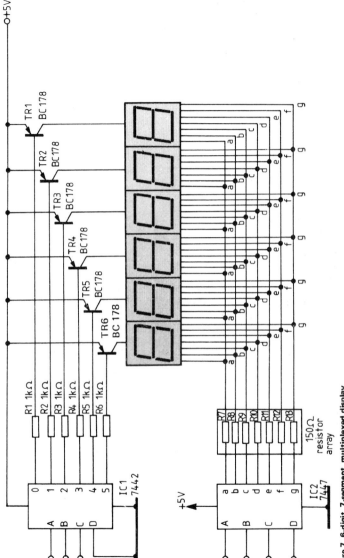

Figure 7 6-digit, 7-segment, multiplexed display

217

```
                    ;Display subroutine (DSPLY)
                    ;Entry: BCD code in A
                    ;A and Y are changed
                    ;
                    DISPNO  =$ØØ8C
                    PAD     =$17ØØ
                         *  =$Ø3D1
                    ;
Ø3D1  48            DSPLY   PHA             ;save A temporarily on the stack
Ø3D2  29 FØ                 AND #$FØ        ;mask off low nibble
Ø3D4  Ø5 8C                 ORA DISPNO      ;and add position code
Ø3D6  E6 8C                 INC DISPNO      ;point to next display position
Ø3D8  8D ØØ 17              STA PAD         ;light up display LED 1
Ø3DB  AØ ØØ                 LDY #$ØØ        ;and hold on
Ø3DD  88            DELAY   DEY             ;for a short time
Ø3DE  DØ FD                 BNE DELAY
Ø3EØ  68                    PLA             ;restore A
Ø3E1  ØA                    ASL A           ;shift low nibble
Ø3E2  ØA                    ASL A           ;into high nibble
Ø3E3  ØA                    ASL A           ;position
Ø3E4  ØA                    ASL A
Ø3E5  Ø5 8C                 ORA DISPNO      ;and add position code
Ø3E7  E6 8C                 INC DISPNO      ;point to next display position
Ø3E9  8D ØØ 17              STA PAD         ;light up display LED 2
Ø3EC  88            DELAY1  DEY             ;for a short time
Ø3ED  DØ FD                 BNE DELAY1
Ø3EF  6Ø                    RTS             ;return to calling program
```

This is an example of the use of **nested subroutines**. The display subroutine
(**DSPLY**) causes a pair of 7-segment LED displays, defined by a position code
stored in location $ØØ8C_{16}$, to be illuminated with BCD characters stored in the
accumulator. This subroutine is called by the multiplexed display subroutine
(**MUXDSP**), and causes the contents of the display buffer $ØØ8D_{16}$, $ØØ8E_{16}$ and
$ØØ8F_{16}$ to illuminate all six LEDs. This relationship is shown in *Figure 8*.

The subroutine (**MUXDSP**) causes all six displays to be illuminated, in turn,
for a short period of time, and must be repeatedly called to **refresh the LEDs**
and provide a continuous display.

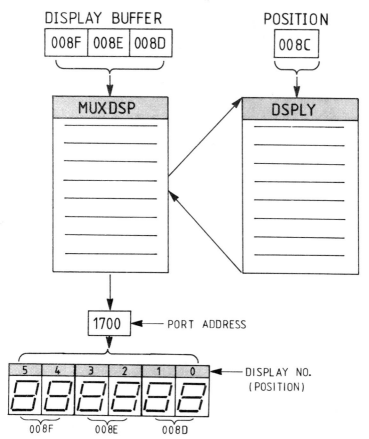

Figure 8

219

Problem 15 Refer to the circuit diagram of a **matrix-type hexadecimal keypad** shown in *Figure 9*.

Write a **6502** machine code subroutine, starting at address Ø3FØ$_{16}$ to scan the keypad until a key closure is detected, and return with the appropriate hexadecimal code in the accumulator. The conents of index registers X and Y should be preserved.

```
                    ;Hexadecimal keypad scan subroutine (INKEY)
                    ;Exit: keycode in A
                    ;X and Y are preserved
                    ;
                    KEYCNT  =$ØØ9Ø
                    PAD     =$17ØØ
                          * =$Ø3FØ
                    ;
Ø3FØ  98    INKEY     TYA
Ø3F1  48              PHA              ;save Y on the stack
Ø3F2  8A              TXA
Ø3F3  48              PHA              ;save X on the stack
Ø3F4  A9 ØØ  SCAN     LDA#Ø            ;clear key counter
Ø3F6  85 9Ø            STA KEYCNT
Ø3F8  A2 F7            LDX #$F7         ;load X with row pattern
Ø3FA  AØ Ø4  NEWROW   LDY #$Ø4         ;use Y as a column counter
Ø3FC  8E ØØ 17         STX PAD          ;send row pattern to keypad
Ø3FF  AD ØØ 17         LDA PAD          ;read columns into A
Ø4Ø2  ØA    SHIFT     ASL A            ;shift column data into carry flag
Ø4Ø3  9Ø ØC            BCC FOUND        ;exit with keycode if key is closed
Ø4Ø5  E6 9Ø            INC KEYCNT       ;advance key counter if key not closed
Ø4Ø7  88              DEY              ;move on to next column
Ø4Ø8  DØ F8            BNE SHIFT        ;see if all four columns checked
Ø4ØA  8A              TXA              ;move X into A to shift it one
Ø4ØB  4A              LSR A            ;place to the right, and
Ø4ØC  AA              TAX              ;put it back, then check for C-flag = Ø
Ø4ØD  BØ EB            BCS NEWROW       ;to see if all 4 row patterns are sent
Ø4ØF  9Ø E3            BCC SCAN         ;no key pressed, so keep on scanning
Ø411  68    FOUND     PLA
Ø412  AA              TAX              ;restore X
Ø413  68              PLA
Ø414  A8              TAY              ;restore Y
Ø415  A5 9Ø            LDA KEYCNT       ;put keycode into A
Ø417  6Ø              RTS              ;return to calling program
```

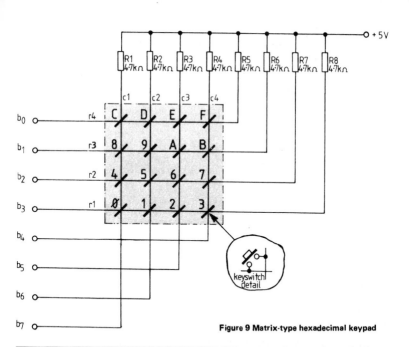

Figure 9 Matrix-type hexadecimal keypad

Problem 16 Show how the important registers in a **Z80** microprocessor may be:
(a) saved **prior to calling** a subroutine, and
(b) restored **after leaving** a subroutine.

(a)

C5	PUSH BC	;save BC on the stack
D5	PUSH DE	;save DE on the stack
E5	PUSH HL	;save HL on the stack
F5	PUSH AF	;save AF on the stack
DD E5	PUSH IX	;save IX on the stack
FD E5	PUSH IY	;save IY on the stack

(b)

FD E1	POP IY	;restore IY from the stack
DD E1	POP IX	;restore IX from the stack
F1	POP AF	;restore AF from the stack
E1	POP HL	;restore HL from the stack
D1	POP DE	;restore DE from the stack
C1	POP BC	;restore BC from the stack

Alternatively, the alternate register set may be used to preserve the contents of most registers by means of the following instructions:

| Ø8 | EX AF,AF′ | ;exchange AF with alternate AF |
| D9 | EXX | ;exchange BC, DE & HL with alternate set |

This may be done, in any order, prior to calling and after leaving a subroutine, provided that the alternate register set is not being used for any other purpose.

Problem 17 A **Z80** machine code subroutine is required to provide a **time delay** of between 1 and 10 seconds, depending upon the value stored in the **HL** register pair prior to calling the subroutine.

Write a suitable subroutine, starting at address $ØEØØ_{16}$ to enable the delay times required to be covered. The values in all other registers except HL should be preserved.

```
                    ;Time delay subroutine (DELAY)
                    ;Entry: HL contains the delay parameter
                    ;Exit: HL=ØØØØ
                    ;All other registers are saved
                    ;
                            ORG ØEØØH
                    ;
ØEØØ  F5      DELAY   PUSH AF         ;save AF on the stack
ØEØ1  C5              PUSH BC         ;save BC on the stack
ØEØ2  Ø6 ØØ           LD B,Ø          ;clear B
ØEØ4  1Ø FE   LOOP    DJNZ LOOP       ;use B as inner loop counter
ØEØ6  2B              DEC HL          ;use HL as outer loop counter
ØEØ7  7C              LD A,H
ØEØ8  B5              OR L            ;set flags
ØEØ9  2Ø F9           JR NZ,LOOP      ;repeat if not counted out
ØEØB  C1              POP BC          ;restore BC
ØEØC  F1              POP AF          ;restore AF
ØEØD  C9              RET             ;return to calling program
```

Table 2

Time delay (seconds)	Value in HL
1	0258
2	04B0
3	0708
4	0960
5	0BB8
6	0E10
7	1068
8	12C0
9	1518
10	1770

Table 2 shows the values required in HL, prior to calling this subroutine, to obtain time delays between 1 and 10 s. These values assume that the program is used in a Z80 based microcomputer with a 2 MHz clock.

Problem 18 A **Z80** machine code subroutine is required to convert the **BCD value** stored in b0–b3 of the accumulator into its equivalent **7-segment code**.

Write a suitable subroutine, starting at address $0E0E_{16}$, to perform the necessary code conversion and return with the segments code in the accumulator.

```
                    ;BCD to 7-segment conversion subroutine (SEG7)
                    ;Entry: BCD data in A
                    ;Exit: segments code in A
                    ;All other registers unchanged
                    ;
                            ORG 0E0EH
                    ;
0E0E  E5      SEG7    PUSH HL          ;save HL on the stack
0E0F  21 17 0E        LD HL,TABLE      ;point HL to segments table
0E12  85              ADD A,L          ;adjust HL to point to
0E13  6F              LD L,A           ;particular code required
0E14  7E              LD A,(HL)        ;and transfer code to A
0E15  E1              POP HL           ;restore HL
0E16  C9              RET              ;return to calling program
0E17  3F      TABLE   DEFB 3FH,06H     ;0, 1
0E18  06
0E19  5B              DEFB 5BH,4FH     ;2, 3
0E1A  4F
0E1B  66              DEFB 66H,6DH     ;4, 5
0E1C  6D
0E1D  7D              DEFB 7DH,07H     ;6, 7
0E1E  07
0E1F  7F              DEFB 7FH,67H     ;8, 9
0E20  67
```

Problem 19 A **Z80** machine code subroutine is required to convert a **packed BCD** value in the accumulator into its **ASCII equivalents** and return these in the HL register pair.

Write a suitable subroutine, starting at address $0E21_{16}$, to perform the necessary code conversion.

```
                      ;BCD to ASCII subroutine (BCDASC)
                      ;Entry: BCD data in A
                      ;Exit: ASCII codes in HL
                      ;Other registers unchanged
                      ;
                              ORG ØE21H
                      ;
ØE21  67      BCDASC  LD H,A        ;temporarily save A in H
ØE22  E6 ØF           AND ØFH       ;mask off high nibble and
ØE24  F6 3Ø           OR 3ØH        ;convert to ASCII
ØE26  6F              LD L,A        ;and put it into L
ØE27  7C              LD A,H        ;restore BCD value in A
ØE28  1F              RRA           ;move high nibble to
ØE29  1F              RRA           ;low nibble position
ØE2A  1F              RRA
ØE2B  1F              RRA
ØE2C  E6 ØF           AND ØFH       ;mask off high nibble
ØE2E  F6 3Ø           OR 3ØH        ;and convert to ASCII
ØE3Ø  67              LD H,A        ;and put it into H
ØE31  C9              RET           ;return to calling program
```

Problem 20 A **Z80** machine code subroutine is required to convert **two ASCII codes** in the HL register pair into **packed BCD** form and return with this in the accumulator.

Write a suitable subroutine, starting at address $ØE32_{16}$, to perform the necessary code conversion.

```
                      ;ASCII to packed BCD subroutine (ASCBCD)
                      ;Entry: ASCII data in HL
                      ;Exit: packed BCD in A
                      ;Other registers unchanged
                      ;
                              ORG ØE32H
                      ;
ØE32  7D      ASCBCD  LD A,L        ;get first ASCII code
ØE33  E6 ØF           AND ØFH       ;strip off ASCII part
ØE35  6F              LD L,A        ;save temporarily in L
ØE36  7C              LD A,H        ;get second ASCII code
ØE37  17              RLA           ;move low nibble to
ØE38  17              RLA           ;high nibble position
ØE39  17              RLA
ØE3A  17              RLA
ØE3B  E6 FØ           AND ØFØH      ;mask off low nibble
ØE3D  B5              OR L          ;combine codes in A
ØE3E  C9              RET           ;return to calling program
```

224

```
                    ;Multiplication subroutine (MULT)
                    ;Entry: multiplier in H, multiplicand in L
                    ;Exit: product in HL
                    ;Other registers unchanged
                    ;
                            ORG ∅E4∅H
                    ;
∅E4∅  C5      MULT    PUSH BC         ;save BC on the stack
∅E41  D5              PUSH DE         ;save DE on the stack
∅E42  5D              LD E,L          ;transfer multiplicand to E
∅E43  2E ∅∅           LD L,∅          ;clear L
∅E45  55              LD D,L          ;clear D
∅E46  ∅6 ∅8           LD B,∅8H        ;use B as loop counter
∅E48  29      SHIFT   ADD HL,HL       ;shift HL to left once
∅E49  3∅ ∅1           JR NC,NOADD     ;skip add if multiplier bit is ∅
∅E4B  19              ADD HL,DE       ;add multiplicand to product
∅E4C  1∅ FA   NOADD   DJNZ SHIFT      ;repeat if not done 8 times
∅E4E  D1              POP DE          ;restore DE
∅E4F  C1              POP BC          ;restore BC
∅E5∅  C9              RET             ;return to calling program
```

```
                    ;Division subroutine (DIVIDE)
                    ;Entry: divisor in H, dividend in L
                    ;Exit: quotient in L, remainder in H
                    ;Other registers unchanged
                    ;
                            ORG ØE51H
                    ;
ØE51  F5      DIVIDE  PUSH AF             ;save AF on the stack
ØE52  D5              PUSH DE             ;save DE on the stack
ØE53  C5              PUSH BC             ;save BC on the stack
ØE54  54              LD D,H              ;transfer divisor to D
ØE55  26 ØØ           LD H,Ø             ;clear H
ØE57  5C              LD E,H              ;clear E
ØE58  Ø6 Ø8           LD B,Ø8H           ;use B as loop counter
ØE5A  29      SHIFT   ADD HL,HL           ;shift HL to left once
ØE5B  ED 52           SBC HL,DE           ;try to divide by D
ØE5D  3Ø Ø1           JR NC,NOADD         ;if possible, skip restore
ØE5F  19              ADD HL,DE           ;restore shifted dividend
ØE6Ø  17      NOADD   RLA                 ;shift bit into quotient
ØE61  10 F7           DJNZ SHIFT          ;repeat if not done 8 times
ØE63  2F              CPL                 ;complement quotient
ØE64  6F              LD L,A              ;and transfer to L
ØE65  C1              POP BC              ;restore BC
ØE66  D1              POP DE              ;restore DE
ØE67  F1              POP AF              ;restore AF
ØE68  C9              RET                 ;return to calling program
```

Problem 23 A **Z80** machine code subroutine is required to convert an **8-bit binary** number in the accumulator into its **decimal** equivalent (BCD) and store this value in the HL register pair.

Write a suitable subroutine, starting at address $ØE69_{16}$.

```
                    ;Binary to BCD conversion subroutine (BIDEC)
                    ;Entry: binary number in A
                    ;Exit: BCD equivalent in HL
                    ;Other registers unchanged
                    ;
                              ORG ØE69H
                    ;
ØE69  C5          BIDEC   PUSH BC         ;save BC on the stack
ØE6A  21 ØØ ØØ            LD HL,Ø         ;clear HL
ØE6D  Ø6 Ø8               LD B,Ø8H        ;use B as loop counter
ØE6F  4F                  LD C,A          ;transfer binary number to C
ØE7Ø  AF                  XOR A           ;clear A
ØE71  CB 11       SHIFT   RL C            ;shift bit into C flag
ØE73  8F                  ADC A,A         ;double A and add carry
ØE74  27                  DAA             ;and convert to BCD
ØE75  CB 14               RL H            ;and shift carry into H
ØE77  1Ø F8               DJNZ SHIFT      ;repeat if not done 8 times
ØE79  6F                  LD L,A          ;transfer lower digits to L
ØE7A  C1                  POP BC          ;restore BC
ØE7B  C9                  RET             ;return to calling program
```

Problem 24 A **Z80** machine code subroutine is required to convert a **decimal (BCD)** number in the accumulator into its **binary** equivalent and store this in register L.

Write a suitable subroutine, starting at address ØE7C₁₆.

```
                    ;BCD to binary subroutine (DECBI)
                    ;Entry: BCD number in A
                    ;Exit: binary equivalent in L
                    ;Other registers unchanged
                    ;
                              ORG ØE7CH
                    ;
ØE7C  C5          DECBI   PUSH BC         ;save BC on the stack
ØE7D  Ø6 Ø8               LD B,Ø8H        ;use B as loop counter
ØE7F  CB 3F       SHIFT   SRL A           ;halve BCD number
ØE81  CB 1D               RR L            ;shift remainder into L
ØE83  CB 5F               BIT 3,A         ;see if there is a carry to b3
ØE85  28 Ø4               JR Z,HCAR       ;skip correction if not
ØE87  D6 Ø3               SUB Ø3H         ;otherwise apply correction
ØE89  18 Ø2               JR ADJ          ;
ØE8B  D6 ØØ       HCAR    SUB Ø           ;dummy subtract for DAA
ØE8D  27          ADJ     DAA             ;decimal adjust number
ØE8E  1Ø EF               DJNZ SHIFT      ;and repeat if not done 8 times
ØE9Ø  C1                  POP BC          ;restore BC
ØE91  C9                  RET             ;return to calling program
```

Write a **Z80** machine code subroutine, starting at address $\emptyset E92_{16}$ to display the BCD contents of registers C, D and E on the LEDs. All other register contents should remain unchanged after using this subroutine.

```
                    ;Multiplexed display subroutine (MUXDSP)
                    ;Entry: BCD codes in registers C, D & E
                    ;This subroutine calls DSPLY subroutine at ØEA7H
                    ;Registers A, F & B are saved
                    ;
            DSPLY   EQU ØEA7H
                    ORG ØE92H
                    ;
ØE92  F5    MUXDSP  PUSH AF         ;save AF registers on the stack
ØE93  C5            PUSH BC         ;save register B on the stack
ØE94  E5            PUSH HL         ;save HL registers on the stack
ØE95  2E ØØ         LD L,Ø          ;L points to right-most LED
ØE97  79            LD A,C          ;get codes for right hand digits
ØE98  CD A7 ØE      CALL DSPLY      ;light up RH pair of digits
ØE9B  7A            LD A,D          ;get codes for centre digits
ØE9C  CD A7 ØE      CALL DSPLY      ;light up centre pair of digits
ØE9F  7B            LD A,E          ;get codes for left hand digits
ØEAØ  CD A7 ØE      CALL DSPLY      ;light up LH pair of digits
ØEA3  E1            POP HL          ;restore HL
ØEA4  C1            POP BC          ;restore B
ØEA5  F1            POP AF          ;restore AF
ØEA6  C9            RET             ;return to calling program
```

This is an example of the use of **nested subroutines**. The display subroutine (**DSPLY**) causes a pair of 7-segment LED displays, defined by a position code stored in register L, to be illuminated with BCD characters whose codes are stored in register A. This subroutine is called by the multiplexed display subroutine (**MUXDSP**), and causes the contents of registers C, D & E to illuminate all six LEDs. This relationship is shown in *Figure 10*.

The subroutine (**MUXDSP**) causes all six displays to be illuminated, in turn, for a short period of time, and must be repeatedly called to **refresh the LEDs** and provide a continuous display.

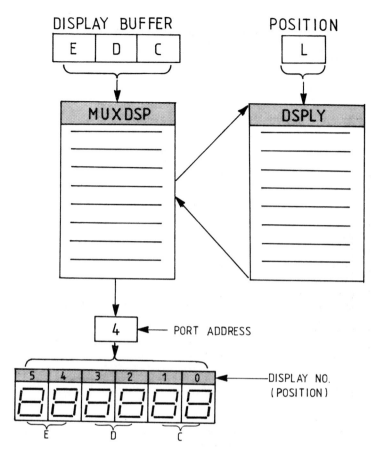

Figure 10

229

```
                    ;Display subroutine (DSPLY)
                    ;Entry: BCD codes in A
                    ;AF, B & L are changed
                    ;
                    PORT  EQU 4
                          ORG ØEA7H
                    ;
ØEA7  F5     DSPLY  PUSH AF          ;save BCD code on the stack
ØEA8  E6 FØ         AND ØFØH         ;mask off least significant BCD code
ØEAA  B5            OR L             ;add LED position code
ØEAB  2C            INC L            ;point L to next LED
ØEAC  D3 Ø4         OUT (PORT),A     ;send code to display unit
ØEAE  Ø6 ØØ         LD B,Ø           ;use B as time delay counter
ØEBØ  1Ø FE  DLY    DJNZ DLY         ;hold LED on for a short time
ØEB2  F1            POP AF           ;recover BCD code from the stack
ØEB3  87            ADD A,A          ;equivalent to shift A to the
ØEB4  87            ADD A,A          ;left by four places
ØEB5  87            ADD A,A
ØEB6  87            ADD A,A
ØEB7  B5            OR L             ;add LED position code
ØEB8  2C            INC L            ;point L to next LED
ØEB9  D3 Ø4         OUT (PORT),A     ;send code to display unit
ØEBB  1Ø FE  DLY1   DJNZ DLY1        ;hold LED on for a short time
ØEBD  C9            RET              ;return to calling program
```

Problem 26 Refer to the circuit diagram of a **matrix-type hexadecimal keypad** shown in *Figure 9*, page 221.

Write a **Z80** machine code subroutine, starting at address $ØEBE_{16}$, to scan the keypad until a key closure is detected, and return with the appropriate hexadecimal code in the accumulator. The contents of all other registers should be preserved.

```
                  ;Hexadecimal keypad scan subroutine (INKEY)
                  ;Exit: keycode in A
                  ;All other registers are saved
                  ;
                  PORT    EQU 5           ;PIO mode 3 (bit mode)
                          ORG ØEBEH
                  ;
ØEBE  C5    INKEY   PUSH BC         ;save BC on the stack
ØEBF  D5            PUSH DE         ;save DE on the stack
ØECØ  ØE Ø5         LD C,PORT       ;access I/O port via C
ØEC2  AF    SCAN    XOR A           ;clear key counter
ØEC3  1E F7         LD E,ØF7H       ;load E with row pattern
ØEC5  Ø6 Ø4  NEWROW LD B,Ø4H        ;use B as a column counter
ØEC7  ED 59         OUT (C),E       ;send row pattern to keypad
ØEC9  ED 5Ø         IN D,(C)        ;read columns into D
ØECB  CB 22  SHIFT   SLA D          ;shift column data into carry flag
ØECD  3Ø Ø9         JR NC,FOUND     ;exit with keycode if key is closed
ØECF  3C            INC A           ;advance key counter if key not closed
ØEDØ  1Ø F9         DJNZ SHIFT      ;see if all four columns checked
ØED2  CB 2B         SRA E           ;shift row bit into carry flag
ØED4  38 EF         JR C,NEWROW     ;to check if all 4 row patterns sent
ØED6  3Ø EA         JR NC,SCAN      ;no key pressed, so keep on scanning
ØED8  D1    FOUND   POP DE          ;restore DE from the stack
ØED9  C1            POP BC          ;restore BC from the stack
ØEDA  C9            RET             ;return to calling program
```

Problem 27 Show how important registers in a **6800** microprocessor may be:

(a) saved **prior to calling** a subroutine; and
(b) restored **after leaving** a subroutine.

(a)

```
36      PSH A       ;save accumulator A on the stack
37      PSH B       ;save accumulator B on the stack
Ø7      TPA
36      PSH A       ;save condition code register on the stack
DF ØØ   STX TEMP    ;save index register X in any zero page location
```

(b)

```
32      PUL A
Ø6      TAP         ;restore condition code register from the stack
33      PUL B       ;restore accumulator B from the stack
32      PUL A       ;restore accumulator A from the stack
DE ØØ   LDX TEMP    ;restore index register X from zero page
```

Note: The 6800 microprocessor has no convenient way of storing index register X on the stack. Therefore this register may be saved prior to calling a subroutine by storing it in any two zero page (address $\emptyset\emptyset\emptyset\emptyset_{16}$ to $\emptyset\emptyset FF_{16}$) locations by means of direct addressing. Index register X may be restored by loading it from the same two zero page locations after leaving the subroutine, and since the stack is not used, the order in which X is saved and restored is unimportant.

Problem 28   A **6800** machine code subroutine is required to provide a **time delay** of between 1 and 10 seconds, depending upon the value stored in **accumulator B** prior to calling the subroutine.

Write a suitable subroutine, starting at address $\emptyset1\emptyset\emptyset_{16}$ to enable the delay times required to be covered. Registers other than accumulator B must be preserved.

```
                    ;Time delay subroutine (DELAY)
                    ;Entry: B contains the delay parameter
                    ;Exit: B = ∅∅
                    ;X is unchanged
                    ;
                    SAVEX  EQU ∅
                           ORG $∅1∅∅
                    ;
∅1∅∅  DF ∅∅   DELAY  STX SAVEX      ;save X register
∅1∅2  CE ∅C ∅∅ LOOP   LDX #$∅C∅∅     ;X is inner loop counter
∅1∅5  ∅9       LOOP1  DEX
∅1∅6  26 FD           BNE LOOP1
∅1∅8  5A              DEC B          ;decrement delay parameter
∅1∅9  26 F7           BNE LOOP       ;repeat until counted out
∅1∅B  DE ∅∅           LDX SAVEX      ;restore X register
∅1∅D  39              RTS            ;return to calling program
```

Table 3

Time delay (seconds)	Value in B
1	19
2	32
3	4B
4	64
5	7D
6	95
7	AF
8	C8
9	E1
10	FA

Table 3 shows the values required in accumulator B, prior to calling this subroutine, to obtain time delays between 1 and 10 seconds. These values assume that the program is used in a 6800 based microcomputer with a 614.4 kHz clock.

```
                    ;BCD to 7-segment conversion routine (SEG7)
                    ;Entry: BCD data in accumulator A
                    ;Exit: segments code in accumulator A
                    ;X and accumulator B are unchanged
                    ;
                    XTEMP  EQU $∅∅∅∅
                           ORG $∅1∅E
                    ;
∅1∅E  DF ∅∅   SEG7    STX XTEMP           ;save index register X
∅11∅  CE ∅1 1B        LDX TABLE–1         ;put X at bottom of table
∅113  ∅8      SEGCNT  INX                 ;point X further up the table
∅114  4A              DEC A               ;by an amount determined by
∅115  2A FC           BNE SEGCNT          ;the BCD number in A
∅117  A6 ∅∅           LDA ∅,X             ;get the segments code in A
∅119  DE ∅∅           LDX XTEMP           ;restore X
∅11B  39              RTS                 ;return to calling program
∅11C  3F      TABLE   FCB $3F,$∅6.$5B     ;∅, 1, 2
∅11D  ∅6
∅11E  5B
∅11F  4F              FCB $4F,$66,$6D     ;3, 4, 5
∅12∅  66
∅121  6D
∅122  7D              FCB $7D,$∅7,$7F     ;6, 7, 8
∅123  ∅7
∅124  7F
∅125  67              FCB $67             ;9
```

```
                    ;BCD to ASCII subroutine (BCDASC)
                    ;Entry: BCD data in accumulator B
                    ;Exit: ASCII codes in $ØØ82 and $ØØ83
                    ;X and accumulator A are unchanged
                    ;
                    ASCLO    EQU $ØØ82
                    ASCHI    EQU $ØØ83
                             ORG $Ø126
                    ;
Ø126   37    BCDASC  PSH B          ;save B on the stack
Ø127   C4 ØF         AND B #$ØF     ;mask off high nibble
Ø129   CA 3Ø         ORA B #$3Ø     ;and convert to ASCII
Ø12B   D7 82         STA B ASCLO    ;and save
Ø12D   33            PUL B          ;restore B
Ø12E   54            LSR B          ;shift low nibble
Ø12F   54            LSR B          ;4 places to the left
Ø13Ø   54            LSR B          ;into high nibble position
Ø131   54            LSR B          ;
Ø132   CA 3Ø         ORA B #$3Ø     ;convert to ASCII
Ø134   D7 83         STA B ASCHI    ;and save
Ø136   39            RTS            ;return to calling program
```

Problem 31 A **6800** machine code subroutine is required to convert **two ASCII codes** in memory locations $ØØ82_{16}$ and $ØØ83_{16}$ into **packed BCD** form.
Write a suitable subroutine, starting at address $Ø137_{16}$, to perform the necessary code conversion and store the packed BCD equivalent in accumulator B. Registers other than accumulator B should be preserved.

```
                    ;ASCII to packed BCD subroutine (ASCBCD)
                    ;Entry: ASCII data in $ØØ82 and $ØØ83
                    ;Exit: packed BCD in accumulator B
                    ;X and accumulator A are unchanged
                    ;
                    ASCLO    EQU $ØØ82
                    ASCHI    EQU $ØØ83
                             ORG $Ø137
                    ;
Ø137   36    ASCBCD  PSH A          ;save accumulator A on the stack
Ø138   D6 82         LDA B ASCLO    ;get low nibble ASCII code
Ø13A   C4 ØF         AND B #$ØF     ;convert ASCII to BCD
Ø13C   96 83         LDA A ASCHI    ;get high nibble ASCII code
Ø13E   48            ASL A          ;convert ASCII to BCD
Ø13F   48            ASL A          ;and shift into high
Ø14Ø   48            ASL A          ;nibble position
Ø141   48            ASL A          ;
Ø142   1B            ABA            ;combine high and low nibbles in A
Ø143   16            TAB            ;and transfer to accumulator B
Ø144   32            PUL A          ;restore accumulator A
Ø145   39            RTS            ;return to calling program
```

A **6800** machine code subroutine is required to **multiply** together two numbers stored in accumulator A and accumulator B, and store the product in memory locations 0084_{16} and 0085_{16}.
Write a suitable subroutine, starting at address 0146_{16}.

```
                    ;Multiplication subroutine (MULT)
                    ;Entry: multiplier in B, multiplicand in A
                    ;Exit: product high in A and in $0084
                    ;      product low in $0085
                    ;      accumulator B = 00
                    ;X is unchanged
                    ;
                    PRODHI  EQU $0084
                    PRODLO  EQU $0085
                            ORG $0146
                    ;
0146   97 84        MULT    STA A PRODHI    ;transfer multiplicand to memory
0148   D7 85                STA B PRODLO    ;transfer multiplier to memory
014A   C6 08                LDA B#$08       ;use B as a loop counter
014C   4F                   CLR A           ;clear product register
014D   46           SHIFT   ROR A           ;shift product/multiplier right
014E   76 00 85             ROR PRODLO      ;one place
0151   24 02                BCC NOADD       ;skip add if multiplier bit is 0
0153   9B 84                ADD A PRODHI    ;add multiplicand to product
0155   5A           NOADD   DEC B           ;decrement loop counter
0156   26 F5                BNE SHIFT       ;repeat if not done 8 times
0158   46                   ROR A           ;final shift
0159   76 00 85             ROR PRODLO      ;of product
015C   97 84                STA A PRODHI    ;
015E   39                   RTS             ;return to calling program
```

Problem 33 A **6800** machine code subroutine is required to **divide** a number in accumulator B by a second number stored in accumulator A.

Write a suitable subroutine, starting at address $\emptyset 15F_{16}$, to divide the two numbers and store the quotient in accumulator B and the remainder in accumulator A.

```
        ;Division subroutine (DIVIDE)
        ;Entry: divisor in accumulator A
        ;       dividend in accumulator B
        ;Exit: quotient in accumulator B
        ; remainder in accumulator A
        ;X is unchanged
        ;
        COUNT   EQU $ØØØ2
        DIVSOR  EQU $ØØ86
                ORG $Ø15F
        ;
Ø15F  97 86      DIVIDE  STA A DIVSOR   ;transfer divisor to memory
Ø161  86 Ø8              LDA A #$Ø8     ;set up a loop counter
Ø163  97 Ø2              STA A COUNT    ;in location $ØØØ2
Ø165  4F                 CLR A          ;clear remainder register
Ø166  59         SHIFT   ROL B          ;shift dividend/quotient
Ø167  49                 ROL A          ;one place to the left
Ø168  9Ø 86              SUB A DIVSOR   ;subtract the divisor
Ø16A  24 Ø2              BCC NOADD      ;do not restore if C=1
Ø16C  9B 86              ADD A DIVSOR   ;divisor too big, so restore
Ø16E  7A ØØ Ø2  NOADD   DEC COUNT      ;decrement loop counter
Ø171  26 F2              BNE SHIFT      ;repeat if not done 8 times
Ø173  59                 ROL B          ;final shift of quotient
Ø174  53                 COM B          ;complement the quotient
Ø175  39                 RTS            ;return to calling program
```

Problem 34 A **6800** machine code subroutine is required to convert an **8-bit binary** number in accumulator A into its **decimal** equivalent (BCD) and store this value in memory locations 0089_{16} and $008A_{16}$.

Write a suitable subroutine, starting at address 0176_{16}, to perform this function.

```
                  ;Binary to BCD conversion subroutine (BIDEC)
                  ;Entry: binary number in accumulator A
                  ;Exit: BCD equivalent in $0089 and $008A
                  ;X and accumulator B are unchanged
                  ;
                  TEMP    EQU $0088
                  BCDLO   EQU $0089
                  BCDHI   EQU $008A
                          ORG $0176
                  ;
0176  37          BIDEC   PSH B          ;save accumulator B on the stack
0177  C6 08               LDA B #$08     ;use accumulator B as a loop counter
0179  97 88               STA A TEMP     ;transfer accumulator A to memory
017B  4F                  CLR A
017C  97 89               STA A BCDLO    ;clear BCDLO
017E  78 00 88    SHIFT   ASL TEMP       ;move each bit into the carry flag
0181  96 89               LDA A BCDLO    ;
0183  99 89               ADC A BCDLO    ;double and add carry
0185  19                  DAA            ;convert to decimal
0186  97 89               STA A BCDLO    ;and put back
0188  79 00 8A            ROL BCDHI      ;shift any carry into BCDHI
018B  5A                  DEC B          ;decrement loop counter
018C  26 F0               BNE SHIFT      ;and repeat if not done 8 times
018E  33                  PUL B          ;restore accumulator B
018F  39                  RTS            ;return to calling program
```

237

```
                  ;BCD to binary subroutine (DECBI)
                  ;Entry: BCD number in accumulator A
                  ;Exit: binary number in $ØØ8B
                  ;X and accumulator B are unchanged
                  ;
                  BINARY EQU $ØØ8B
                         ORG $Ø19Ø
                  ;
Ø19Ø   37         DECBI   PSH B         ;save accumulator B on the stack
Ø191   C6 Ø8              LDA B #$Ø8    ;use accumulator B as a loop counter
Ø193   44         SHIFT   LSR A         ;divide accumulator A by 2
Ø194   76 ØØ 8B           ROR BINARY    ;shift remainder into BINARY
Ø197   85 Ø8              BIT A #$Ø8    ;see if A contains illegal BCD
Ø199   27 Ø3              BEQ HCAR      ;skip correction if not
Ø19B   8Ø Ø3              SUB A # $Ø3   ;otherwise apply correction
Ø19D   19                 DAA           ;decimal adjust
Ø19E   5A         HCAR    DEC B         ;decrement loop counter
Ø19F   26 F2              BNE SHIFT     ;repeat if not done 8 times
Ø1A1   33                 PUL B         ;restore accumulator B
Ø1A2   39                 RTS           ;return to calling program
```

```
                    ;Multiplexed display subroutine (MUXDSP)
                    ;Entry: BCD codes in $ØØ8D, $ØØ8E & $ØØ8F
                    ;A, B & X are preserved
                    ;This subroutine calls DSPLY subroutine at $Ø1BD
                    ;
                    TEMP    EQU $ØØØØ
                    DISPNO  EQU $ØØ8C
                    DISBUF  EQU $ØØ8D
                            ORG $Ø1A3
                    ;
Ø1A3 DF ØØ          MUXDSP STX TEMP       ;save X
Ø1A5 36                    PSH A          ;save A on the stack
Ø1A6 37                    PSH B          ;save B on the stack
Ø1A7 7F ØØ 8C              CLR DISPNO     ;initialise display counter
Ø1AA CE ØØ 8D              LDX #DISBUF    ;point X to buffer
Ø1AD A6 ØØ          DISPCD LDA A,X        ;get code from buffer
Ø1AF BD Ø1 BD              JSR DSPLY      ;light up one pair of digits
Ø1B2 Ø8                    INX            ;point to next position in buffer
Ø1B3 8C ØØ 9Ø              CPX DISBUF+3   ;check for buffer end
Ø1B6 26 F5                 BNE DISPCD     ;continue if not at buffer end
Ø1B8 33                    PUL B          ;restore B from the stack
Ø1B9 32                    PUL A          ;restore A from the stack
Ø1BA DE ØØ                 LDX TEMP       ;restore X
Ø1BC 39                    RTS            ;return to calling program
```

```
                    ;Display subroutine (DSPLY)
                    ;Entry: BCD code in A
                    ;A and B are changed
                    ;
                    DISPNO EQU $ØØ8C
                    DRA    EQU $8ØØ4
                           ORG $Ø1BD
Ø1BD 36             DSPLY  PSH A          ;save A temporarily on the stack
Ø1BE 84 FØ                 AND A #$FØ     ;mask of low nibble
Ø1CØ 9A 8C                 ORA A DISPNO   ;and add position code
Ø1C2 7C ØØ 8C              INC DISPNO     ; point to next display position
Ø1C5 B7 8Ø Ø4              STA A DRA      ;light up display LED 1
Ø1C8 5F                    CLR B          ;and hold on
Ø1C9 5A             DELAY  DEC B          ;for a short time
Ø1CA 26 FD                 BNE DELAY      ;
Ø1CC 32                    PUL A          ;restore A
Ø1CD 48                    ASL A          ;shift low nibble
Ø1CE 48                    ASL A          ;into high nibble
Ø1CF 48                    ASL A          ;position
Ø1DØ 48                    ASL A
Ø1D1 9A 8C                 ORA A DISPNO   ;and add position code
Ø1D3 7C ØØ 8C              INC DISPNO     ;point to next display position
Ø1D6 B7 8Ø Ø4              STA A DRA      ;light up display LED 2
Ø1D9 5A             DELAY1 DEC B          ;for a short time
Ø1DA 26 FD                 BNE DELAY1
Ø1DC 39                    RTS            ;return to calling program
```

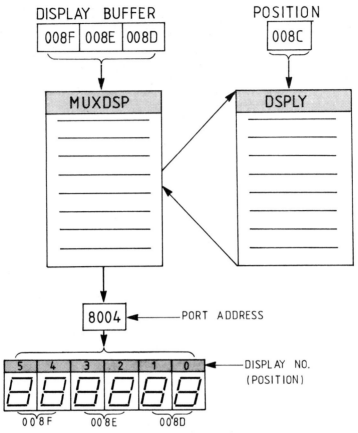

Figure 11

This is an example of the use of **nested subroutines**. The display subroutine **(DSPLY)** causes a pair of 7-segment LED displays, defined by a position code stored in location $\emptyset\emptyset 8C_{16}$, to be illuminated with BCD characters stored in accumulator A. This subroutine is called by the multiplexed display subroutine **(MUXDSP)**, and causes the contents of the display buffer $\emptyset\emptyset 8D_{16}$, $\emptyset\emptyset 8E_{16}$ and $\emptyset\emptyset 8F_{16}$ to illuminate all six LEDs. This relationship is shown in *Figure 11*.

The subroutine **(MUXDSP)** causes all six displays to be illuminated, in turn, for a short period of time, and must be repeatedly called to **refresh the LEDs** and provide a continuous display.

Problem 37 Refer to the circuit diagram of a **matrix-type hexadecimal keypad** shown in *Figure 9*, page 221.

Write a **6800** machine code subroutine, starting at address $\emptyset1DD_{16}$ to scan the keypad until a key closure is detected, and return with the appropriate hexadecimal code in accumulator A. The contents of accumulator B and index register X should remain unchanged.

```
                    ;Hexadecimal keypad scan subroutine (INKEY)
                    ;Exit: keycode in accumulator A
                    ;B and X are unchanged
                    ;
                    KEYCNT   EQU $ØØ9Ø
                    COLCNT   EQU $ØØ91
                    DRA      EQU $8ØØ4
                    ;
Ø1DD 37     INKEY   PUSH B             ;save B on the stack
Ø1DE 7F ØØ 9Ø  SCAN CLR KEYCNT        ;clear key counter
Ø1E1 C6 F7          LDA B#$F7          ;load B with row pattern
Ø1E3 86 Ø4  NEWROW  LDA A#$Ø4          ;
Ø1E5 97 91          STA A COLCNT       ;initialise the column counter
Ø1E7 F7 8Ø Ø4       STA B DRA          ;send row pattern to keypad
Ø1EA B6 8Ø Ø4       LDA A DRA          ;read columns into A
Ø1ED 48     SHIFT   ASL A              ;shift column data into carry flag
Ø1EE 24 ØD          BCC FOUND          ;exit with keycode if key is closed
Ø1FØ 7C ØØ 9Ø       INC KEYCOUNT       ;advance key counter if key not closed
Ø1F3 7A ØØ91        DEC COLCNT         ;move on to next column
Ø1F6 26 F5          BNE SHIFT          ;all four columns checked?
Ø1F8 54             LSR B              ;shift row pattern to right
Ø1F9 25 E8          BCS NEWROW         ;repeat if all four rows not tested
Ø1FB 2Ø E1          BRA SCAN           ;no key pressed, so keep on scanning
Ø1FD 33     FOUND   PUL B              ;restore B
Ø1FE 96 9Ø          LDA A KEYCNT       ;put keycode into A
Ø2ØØ 39             RTS                ;return to calling program
```

C FURTHER PROBLEMS ON SUBROUTINES AND THE STACK

(a) SHORT ANSWER PROBLEMS

1 Sequences of instructions which perform general purpose functions are known as ....

2 A subroutine is a .... which may be called from ....
.... in a program, and may be used ....

3 In order to make use of a subroutine, a .... instruction must be executed.

4 In order to leave a subroutine and go back to the calling program, a ....
instruction must be executed.

5 On advantage of using a subroutine in a program is that it ....
....

6 One disadvantage of using a subroutine in a program is that it ....
....

7 When a subroutine is used, it is necessary to save a ....
so that the calling program may resume execution from the correct point.

8 A 'stack' is a form of .... register.

9 A stack may be used to store .... when using subroutines.

10 In order to maintain a software stack in RAM, a microprocessor has a ....
....

11 A 'nested subroutine' is an arrangement in a program where a subroutine ....
....

12 The depth to which subroutines may be nested in a program is determined by
the ....

13 Data may be saved on the stack by using a .... instruction.

14 Data may be retrieved from the stack by using a .... instruction.

15 The transfer of data from a program to a subroutine or from a subroutine to a
program is known as ....

(b) CONVENTIONAL PROBLEMS

1 Explain the operation of the stack for each of the following microprocessors:
(i) 6502; (ii) Z80; (iii) 6800.

2 With the aid of a suitable sketch, explain why it is important to initialise the
stack pointer of a microprocessor before subroutines are called into use by a
program.

3 A 6502 machine code subroutine is required to append an odd parity bith (bit 7)
to ASCII data stored in the accumulator (i.e. bit 7 must be such as to cause the
total number of logical 1's in the accumulator to be odd for all possible ASCII
codes). Write an appropriate subroutine, starting at any convenient address, to
perform the required function, ensuring that index registers X and Y are not
corrupted.

4 A 6502 machine code subroutine is required to produce a 2 kHz tone, duration 1
second, in an audio transducer connected to bit $\emptyset$ of an I/O port. Write a
suitable subroutine, starting at any convenient address, to perform this
function, ensuring that index registers X and Y are not corrupted.

5 A 6502 machine code subroutine is required to read the keypad shown in *Figure 9*
(page 221), and generate the appropriate ASCII code for each key press, and
return with this in the accumulator. Write a suitable subroutine, starting at any
convenient address, to perform this function, ensuring that index registers X
and Y are not corrupted.

6 A 6502 machine code subroutine is required to convert a binary number in the
accumulator into its equivalent ASCII codes, check that the codes generated
are within the accepted range for hexadecimal notation, and store the codes in

memory locations $\emptyset\emptyset82_{16}$ and $\emptyset\emptyset83_{16}$. If the codes are outside the range for $\emptyset$–9 and A–F, the subroutine must return with the value FF_{16} in the accumulator, otherwise the subroutine returns with $\emptyset\emptyset$ in the accumulator, Write a suitable subroutine, starting at any convenient address, to perform this function.

7 A 6502 machine code subroutine is required to convert the ASCII codes in memory locations $\emptyset\emptyset82_{16}$ and $\emptyset\emptyset83_{16}$ into packed BCD form and return with this in the accumulator. A check must be made to determine that the code stored in the accumulator is valid BCD, and if not, the subroutine must return with FF_{16} in the accumulator. Write a suitable subroutine, starting at any convenient address to perform this function.

8 Serial data of the type shown in *Figure 12* is connected to bit $\emptyset$ of an I/O port. Assuming that appropriate time delay subroutines are available already, write a suitable 6502 machine code subroutine, starting at any convenient address, to read in the serial data, bit by bit into the accumulator, and return when all 8 data bits have been received.

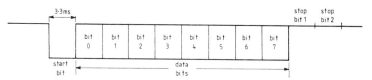

Figure 12

9 A 6502 machine code subroutine is required to transform data in the accumulator into serial form of the type shown in *Figure 12* and send this to bit 1 of an I/O port. Assuming that appropriate time delay subroutines are available already, write a suitable subroutine, starting at any convenient address, to perform the conversion from parallel to serial data form.

10 A Z80 machine code subroutine is required to append an odd parity bit (bit 7) to ASCII data stored in the accumulator (i.e. bit 7 must be such as to cause the total number of logical 1's in the accumulator to be odd for all possible ASCII codes). Write an appropriate subroutine, starting at any convenient address, to perform the required function, ensuring that all other registers are preserved.

11 A Z80 machine code subroutine is required to produce a 2 kHz tone, for a duration of one second, in an audio transducer connected to bit $\emptyset$ of an I/O port. Write a suitable subroutine, starting at any convenient address, to perform this function, ensuring that all registers used in the subroutine are preserved.

12 A Z80 machine code subroutine is required to read the keypad shown in *Figure 9* (page 221), and generate the appropriate ASCII code for each key press, and return with this in the accumulator. Write a suitable subroutine, starting at any convenient address, to perform this function, ensuring that all other registers are preserved.

13 A Z80 machine code subroutine is required to convert a binary number in the accumulator into its equivalent ASCII codes, check that the codes generated are within the accepted range for hexadecimal notation, and store the codes in the HL register pair. If the codes are outside the range for 0–9 and A–F, the

subroutine must return with the value FF_{16} in the accumulator, otherwise the subroutine returns with $\emptyset\emptyset$ in the accumulator. Write a suitable subroutine, starting at any convenient address, to perform this function.

14 A Z80 machine code subroutine is required to convert the ASCII codes stored in the HL register pair into packed BCD form and return with this in the accumulator. A check must be made to determine that the code stored in the accumulator is valid BCD, and if not, the subroutine must return with FF_{16} in the accumulator. Write a suitable subroutine, starting at any convenient address, to perform this function.

15 Serial data of the type shown in *Figure 12* (page 243), is connected to bit $\emptyset$ of an I/O port. Assuming that appropriate time delay subroutines are available already, write a suitable Z80 machine code subroutine, starting at any convenient address, to read in the serial data, bit by bit into the accumulator, and return when all 8 bits have been received.

16 A Z80 machine code subroutine is required to transform data in the accumulator into serial form of the type shown in *Figure 12* and send this to bit 1 of an I/O port. Assuming that appropriate time delay subroutines are available already, write a suitable subroutines, starting at any convenient address, to perform the conversion from parallel to serial data form.

17 A 6800 machine code subroutine is required to append an odd parity bit (bit7) to ASCII data stored in accumulator A (i.e. bit 7 must be such as to cause the total number of logical 1's in accumulator A to be odd for all possible ASCII codes). Write an appropriate subroutine, starting at any convenient address, to perform the required function, ensuring that all other registers are preserved.

18 A 6800 machine code subroutine is required to produce a 2 kHz tone, for a duration of one second, in an audio transducer connected to bit $\emptyset$ of an I/O port. Write a suitable subroutine, starting at any convenient address, to perform this function, ensuring that all registers are preserved.

19 A 6800 machine code subroutine is required to read the keypad shown in *Figure 9* (page 221), and generate the appropriate ASCII code for each key press, and return with this in accumulator A. Write a suitable subroutine, starting at any convenient address, to perform this function, ensuring that all other registers are preserved.

20 A 6800 machine code subroutine is required to convert a binary number in accumulator A into its equivalent ASCII codes, check that the codes generated are within the accepted range for hexadecimal notation, and store the codes in memory locations $\emptyset\emptyset82_{16}$ and $\emptyset\emptyset83_{16}$. If the codes are outside the range for $\emptyset$–9 and A–F, the subroutine must return with the value FF_{16} in accumulator A, otherwise the subroutine returns with $\emptyset\emptyset$ in accumulator A. Write a suitable subroutine, starting at any convenient address, to perform this function.

21 A 6800 machine code subroutine is required to convert the ASCII codes in memory locations $\emptyset\emptyset82_6$ and $\emptyset\emptyset83_{16}$ into packed BCD form and return with this in accumulator A. A check must be made to determine that the code stored in accumulator A is valid BCD, and if not, the subroutine must return with FF_{16} in accumulator A. Write a suitable subroutine, starting at any convenient address to perform this function.

22 Serial data of the type shown in *Figure 12* (page 243) is connected to bit Ø of an I/O port. Assuming that appropriate time delay subroutines are available already, write a suitable 6800 machine code subroutine, starting at any convenient address, to read in the serial data, bit by bit into accumulator A, and return when all 8 data bits have been received.

23 A 6800 machine code subroutine is required to transform data in accumulator A into serial form of the type shown in *Figure 12* (page 243), and send this to bit 1 of an I/O port. Assuming that appropriate time delay subroutines are availble already, write a suitable subroutine, starting at any convenient address, to perform the conversion from parallel to serial form.

Appendix A: Logic functions

A MAIN POINTS CONCERNED WITH LOGIC FUNCTIONS

1   In systems using digital techniques, a need frequently arises to manipulate data, perform arithmetic or make decisions based upon prevailing conditions. This is true for both hardware and software oriented systems, and in order to perform such operations, logical processes are carried out by the use of hardware or software **'logic functions'**. Hardware logic functions are implemented by the use of electronic circuits called **'logic gates'**.

2   Logic functions may be represented by means of combinations of electrical switches, the operation of which are defined in *Figure 1*. A logic function has one or more inputs (independent variables) which may be **true** (logical 1) or **false** (logical $\emptyset$) at any instant in time.

   The output of a particular logic function depends upon the states of each of its inputs and is true if the input conditions specified by the function are also true. One convenient method of describing a logic function is by the use of a **'truth table'** in which the output state for all possible input combinations is tabulated (see *Table 1*).

3   The most common logic functions are:
   (i) NOT, (ii) AND, (iii) INCLUSIVE OR, and (iv) EXCLUSIVE OR.
   Combinations of these give two further logic functions:
   (iv) NAND (NOT–AND), and (v) NOR (NOT–OR).

Table 1

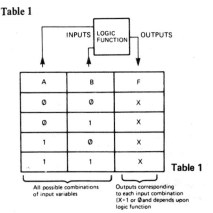

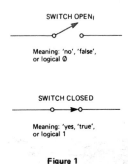

Figure 1

The behaviour of each of these functions may be summarised as follows:

(i) **NOT function** This is the simplest of the logic functions, having only one input, and it is used to invert or complement logic levels. The symbols commonly used to represent a NOT gate are shown in *Figure 2(a)* and the corresponding truth table is shown in *Figure 2(b)*.

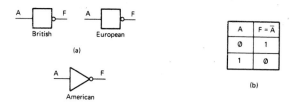

A	$F = \bar{A}$
0	1
1	0

(b)

Figure 2

(ii) **AND function** A two-input AND function is true if both its inputs are true, i.e. both its inputs must be at logical 1 to obtain an output of logical 1. The symbols commonly used to represent an AND gate are shown in *Figure 3(a)* and the corresponding truth table is shown in *Figure 3(b)*. In terms of electrical switching circuits, the AND function is equivalent to two switches connected in series, see *Figure 3(c)*.

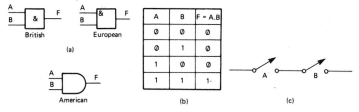

A	B	$F = A.B$
0	0	0
0	1	0
1	0	0
1	1	1

(b)

Figure 3

(iii) **INCLUSIVE OR function** The inclusive OR of two inputs is true if either (or both) of the inputs is true, i.e. either one of its inputs, or both, must be at logical 1 to obtain an output of logical 1. The symbols commonly used to represent an inclusive OR gate are shown in *Figure 4(a)* and the corresponding truth table is shown in *Figure 4(b)*. In terms of electrical switching circuits, the inclusive OR function is equivalent to two switches connected in parallel, see *Figure 4(c)*.

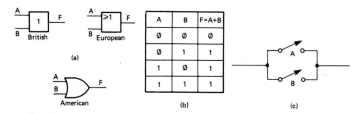

A	B	$F = A + B$
0	0	0
0	1	1
1	0	1
1	1	1

(b)

Figure 4

(iv) **EXCLUSIVE OR function** An exclusive OR function is true if either of its inputs is true, but false if both inputs are true, i.e. either of its inputs (but not both) must be at logical 1 to obtain an output of logical 1. An exclusive OR function is commonly represented by the symbol $\oplus$. Symbols commonly used to represent an exclusive OR gate are shown in *Figure 5(a)* and the corresponding truth table is shown in *Figure 5(b)*.

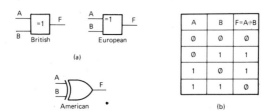

(a)

A	B	F=A⊕B
$\emptyset$	$\emptyset$	$\emptyset$
$\emptyset$	1	1
1	$\emptyset$	1
1	1	$\emptyset$

(b)

Figure 5

(v) **NAND function** A two-input NAND function is true except when both its inputs are true, i.e. an output of logical 1 is obtained except for when its inputs are both at logical 1. The symbols commonly used to represent a NAND gate are shown in *Figure 6(a)* and the corresponding truth table is shown in *Figure 6(b*

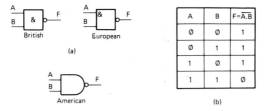

(a)

A	B	F=$\overline{A.B}$
$\emptyset$	$\emptyset$	1
$\emptyset$	1	1
1	$\emptyset$	1
1	1	$\emptyset$

(b)

Figure 6

(vi) **NOR function** A two input NOR function is true only when its inputs are both false, i.e. an output of logical 1 is obtained only when both inputs are at logical $\emptyset$. The symbols commonly used to represent a NOR gate are shown in *Figure 7(a)* and the corresponding truth table is shown in *Figure 7(b)*.

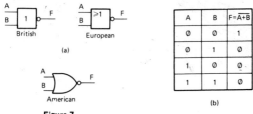

(a)

A	B	F=$\overline{A+B}$
$\emptyset$	$\emptyset$	1
$\emptyset$	1	$\emptyset$
1	$\emptyset$	$\emptyset$
1	1	$\emptyset$

(b)

Figure 7

4   For logic design purposes, a special form of mathematics is frequently used which is known as 'Boolean algebra'. A detailed explanation of the rules of Boolean algebra is outside the scope of this appendix, however, a basic knowledge of the notation used may form a useful aid to understanding technical data associated with microprocessor systems. A summary of Boolean notation is as follows:

(i)   Variables are denoted by using letters of the alphabet with or without subscripts,

      e.g. $A, B, C, \ldots \ldots \ldots \ldots$,

                  $A_0, A_1, A_2, \ldots \ldots \ldots \ldots$,

      or $CS_1, CS_2, CS_3, \ldots \ldots \ldots \ldots$.

(ii) Variables may be used in their true or complemented (not) form. The symbol for a 'NOT' function is a bar, and this is written above all variables expressed in complemented form. Thus, if $A=1$, $\overline{A}=\emptyset$, and if $CS=1$, $\overline{CS}=\emptyset$.

(iii) Where appropriate, the logic state required to cause a particular action is written beside each input/output of an integrated circuit. Examples of this include:

    (a) **CS** (chip select): this indicates an **'active high'** input and a logical 1 is required for chip selection,

    (b) **$\overline{RST}$** (not reset): this indicates an **'active low'** input and a logical $\emptyset$ is required to cause a reset to take place,

    (c) **$R/\overline{W}$** (read/not write): this indicates that a logical 1 applied to this input causes the chip to be in a condition for 'read' operations to take place, but a logical $\emptyset$ conditions the chip for 'write' operations.

(iv) The symbol used to denote a logical AND operation is '•' (since the result of a logical AND is similar to that obtained with an arithmetic multiplication). Thus **A.B** means '**A** logically ANDed with **B**'.

(v)   The symbol used to denote a logical OR operation is '+' (since the result of a logical OR is similar to that obtained with an arithmetic addition). Thus A+B means '**A** logically ORed with **B**'.

(vi) The symbol used to denote an exclusive OR operation is '$\oplus$'. Thus $A \oplus B$ means '**A** exclusively ORed with **B**'. Logically this may also be written as $A.\overline{B} + \overline{A}.B$.

Appendix B;
Instruction sets for
the 6502, Z80 and
6800 microprocessors

The following microprocessor instruction sets are included in this appendix

MS 6502 (pages 251 and 252) from MOS Technology Inc.

Z 80 (pages 253 to 263) from Mostek UK Ltd.

MC 6800 (pages 264 to 265) from Motorola Semiconductor Products Inc.

The author and publishers would like to thank the manufacturers concerned for their permission to publish this information.

THE MCS 6502 INSTRUCTION SET

MNEMONIC	OPERATION	IMMEDIATE OP/N/#	ABSOLUTE OP/N/#	ZERO PAGE OP/N/#	ACCUM OP/N/#	IMPLIED OP/N/#	(IND,X) OP/N/#	(IND),Y OP/N/#	Z.PAGE,X OP/N/#	ABS,X OP/N/#	ABS,Y OP/N/#	RELATIVE OP/N/#	INDIRECT OP/N/#	Z,PAGE,Y OP/N/#	N	Z	C	I	D	V
ADC	A+M+C → A (4)(1)	69 2 2	6D 4 3	65 3 2			61 6 2	71 5 2	75 4 2	7D 4 3	79 4 3				✓	✓	✓	–	–	✓
AND	A∧M→A (1)	29 2 2	2D 4 3	25 3 2			21 6 2	31 5 2	35 4 2	3D 4 3	39 4 3				✓	✓	–	–	–	–
ASL	C←□←0		0E 6 3	06 5 2	0A 2 1				16 6 2	1E 7 3					✓	✓	✓	–	–	–
BCC	BRANCH ON C=0 (2)											90 2 2			–	–	–	–	–	–
BCS	BRANCH ON C=1 (2)											B0 2 2			–	–	–	–	–	–
BEQ	BRANCH ON Z=1 (2)											F0 2 2			–	–	–	–	–	–
BIT	A∧M		2C 4 3	24 3 2											M7	✓	–	–	–	M6
BMI	BRANCH ON N=1 (2)											30 2 2			–	–	–	–	–	–
BNE	BRANCH ON Z=0 (2)											D0 2 2			–	–	–	–	–	–
BPL	BRANCH ON N=0 (2)											10 2 2			–	–	–	–	–	–
BRK	(See Fig. 1)					00 7 1									–	–	–	1	–	–
BVC	BRANCH ON V=0 (2)											50 2 2			–	–	–	–	–	–
BVS	BRANCH ON V=1 (2)											70 2 2			–	–	–	–	–	–
CLC	0→C					18 2 1									–	–	0	–	–	–
CLD	0→D					D8 2 1									–	–	–	–	0	–
CLI	0→I					58 2 1									–	–	–	0	–	–
CLV	0→V					B8 2 1									–	–	–	–	–	0
CMP	A-M	C9 2 2	CD 4 3	C5 3 2			C1 6 2	D1 5 2	D5 4 2	DD 4 3	D9 4 3				✓	✓	✓	–	–	–
CPX	X-M	E0 2 2	EC 4 3	E4 3 2											✓	✓	✓	–	–	–
CPY	Y-M	C0 2 2	CC 4 3	C4 3 2											✓	✓	✓	–	–	–
DEC	M-1→M		CE 6 3	C6 5 2					D6 6 2	DE 7 3					✓	✓	–	–	–	–
DEX	X-1→X					CA 2 1									✓	✓	–	–	–	–
DEY	Y-1→Y					88 2 1									✓	✓	–	–	–	–
EOR	A∀M→A (1)	49 2 2	4D 4 3	45 3 2			41 6 2	51 5 2	55 4 2	5D 4 3	59 4 3				✓	✓	–	–	–	–
INC	M+1→M		EE 6 3	E6 5 2					F6 6 2	FE 7 3					✓	✓	–	–	–	–
INX	X+1→X					E8 2 1									✓	✓	–	–	–	–
INY	Y+1→Y					C8 2 1									✓	✓	–	–	–	–
JMP	JUMP TO NEW LOC.		4C 3 3										6C 5 3		–	–	–	–	–	–
JSR	(See Fig. 2) JUMP SUB		20 6 3												–	–	–	–	–	–
LDA	M → A (1)	A9 2 2	AD 4 3	A5 3 2			A1 6 2	B1 5 2	B5 4 2	BD 4 3	B9 4 3				✓	✓	–	–	–	–

251

THE MS 6502 INSTRUCTION SET

MNEMONIC	OPERATION	IMMEDIATE OP	N	#	ABSOLUTE OP	N	#	ZERO PAGE OP	N	#	ACCUM OP	N	#	IMPLIED OP	N	#	(IND,X) OP	N	#	(IND),Y OP	N	#	Z,PAGE,X OP	N	#	ABS,X OP	N	#	ABS,Y OP	N	#	RELATIVE OP	N	#	INDIRECT OP	N	#	Z,PAGE,Y OP	N	#	N	Z	C	I	D	V
LDX	M → X	A2	2	2	AE	4	3	A6	3	2																													B6	4	2	✓	✓	—	—	—	—
LDY	M → Y	A0	2	2	AC	4	3	A4	3	2													B4	4	2	BC	4	3													✓	✓	—	—	—	—
LSR	0→[ ]→C				4E	6	3	46	5	2	4A	2	1										56	6	2	5E	7	3													0	✓	✓	—	—	—
NOP	NO OPERATION													EA	2	1																							—	—	—	—	—	—
ORA	A∨M → A	09	2	2	0D	4	3	05	3	2							01	6	2	11	5	2	15	4	2	1D	4	3	19	4	3										✓	✓	—	—	—	—
PHA	A → Ms, S-1 → S													48	3	1																							—	—	—	—	—	—
PHP	P → Ms, S-1 → S													08	3	1																							—	—	—	—	—	—
PLA	S+1 → S, Ms → A													68	4	1																							✓	✓	—	—	—	—
PLP	S+1 → S, Ms → P													28	4	1																							(RESTORED)					
ROL	[ ]				2E	6	3	26	5	2	2A	2	1										36	6	2	3E	7	3													✓	✓	✓	—	—	—
ROR	[ ]				6E	6	3	66	5	2	6A	2	1										76	6	2	7E	7	3													✓	✓	✓	—	—	✓
RTI	(See Fig 1) RTRN INT													40	6	1																							(RESTORED)					
RTS	(See Fig 2) RTRN SUB													60	6	1																							—	—	—	—	—	—
SBC	A-M-C̄ → A	(1)E9	2	2	ED	4	3	E5	3	2							E1	6	2	F1	5	2	F5	4	2	FD	4	3	F9	4	3										✓	✓	✓(3)	—	—	✓
SEC	1 → C													38	2	1																							—	—	1	—	—	—
SED	1 → D													F8	2	1																							—	—	—	—	1	—
SEI	1 → I													78	2	1																							—	—	—	1	—	—
STA	A → M				8D	4	3	85	3	2							81	6	2	91	6	2	95	4	2	9D	5	3	99	5	3										—	—	—	—	—	—
STX	X → M				8E	4	3	86	3	2																												96	4	2	—	—	—	—	—	—
STY	Y → M				8C	4	3	84	3	2													94	4	2																—	—	—	—	—	—
TAX	A → X													AA	2	1																							✓	✓	—	—	—	—
TAY	A → Y													A8	2	1																							✓	✓	—	—	—	—
TSX	S → X													BA	2	1																							✓	✓	—	—	—	—
TXA	X → A													8A	2	1																							✓	✓	—	—	—	—
TXS	X → S													9A	2	1																							—	—	—	—	—	—
TYA	Y → A													98	2	1																							✓	✓	—	—	—	—

X   INDEX X
Y   INDEX Y
A   ACCUMULATOR
M   MEMORY PER EFFECTIVE ADDRESS
Ms MEMORY PER STACK POINTER

+   ADD
-   SUBTRACT
∧   AND
∨   OR
⊻   EXCLUSIVE OR

✓   MODIFIED
—   NOT MODIFIED
M₇ MEMORY BIT 7
M₆ MEMORY BIT 6

N   NO CYCLES
NO BYTES

(1) ADD 1 TO "N" IF PAGE BOUNDARY IS CROSSED
(2) ADD 1 TO "N" IF BRANCH OCCURS TO SAME PAGE.
    ADD 2 TO "N" IF BRANCH OCCURS TO DIFFERENT PAGE.
(3) CARRY NOT = BORROW.
(4) IF IN DECIMAL MODE Z FLAG IS INVALID
    ACCUMULATOR MUST BE CHECKED FOR ZERO RESULT.

8-BIT LOAD GROUP

Mnemonic	Symbolic Operation	S	Z		H		P/V	N	C	76 543 210	Hex	No. of Bytes	No. of M Cycles	No. of T States	Comments	
LD r, s	r ← s	•	•	X	•	X	•	•	•	01 r s		1	1	4	r, s	Reg.
LD r, n	r ← n	•	•	X	•	X	•	•	•	00 r 110		2	2	7	000	B
										← n →					001	C
LD r, (HL)	r ← (HL)	•	•	X	•	X	•	•	•	01 r 110		1	2	7	010	D
LD r, (IX+d)	r ← (IX+d)	•	•	X	•	X	•	•	•	11 011 101	DD	3	5	19	011	E
										01 r 110					100	H
										← d →					101	L
LD r, (IY+d)	r ← (IY+d)	•	•	X	•	X	•	•	•	11 111 101	FD	3	5	19	111	A
										01 r 110						
										← d →						
LD (HL), r	(HL) ← r	•	•	X	•	X	•	•	•	01 110 r		1	2	7		
LD (IX+d), r	(IX+d) ← r	•	•	X	•	X	•	•	•	11 011 101	DD	3	5	19		
										01 110 r						
										← d →						
LD (IY+d), r	(IY+d) ← r	•	•	X	•	X	•	•	•	11 111 101	FD	3	5	19		
										01 110 r						
										← d →						
LD (HL), n	(HL) ← n	•	•	X	•	X	•	•	•	00 110 110	36	2	3	10		
										← n →						
LD (IX+d), n	(IX+d) ← n	•	•	X	•	X	•	•	•	11 011 101	DD	4	5	19		
										00 110 110	36					
										← d →						
										← n →						
LD (IY+d), n	(IY+d) ← n	•	•	X	•	X	•	•	•	11 111 101	FD	4	5	19		
										00 110 110	36					
										← d →						
										← n →						
LD A, (BC)	A ← (BC)	•	•	X	•	X	•	•	•	00 001 010	0A	1	2	7		
LD A, (DE)	A ← (DE)	•	•	X	•	X	•	•	•	00 011 010	1A	1	2	7		
LD A, (nn)	A ← (nn)	•	•	X	•	X	•	•	•	00 111 010	3A	3	4	13		
										← n →						
										← n →						
LD (BC), A	(BC) ← A	•	•	X	•	X	•	•	•	00 000 010	02	1	2	7		
LD (DE), A	(DE) ← A	•	•	X	•	X	•	•	•	00 010 010	12	1	2	7		
LD (nn), A	(nn) ← A	•	•	X	•	X	•	•	•	00 110 010	32	3	4	13		
										← n →						
										← n →						
LD A, I	A ← I	↕	↕	X	0	X	IFF	0	•	11 101 101	ED	2	2	9		
										01 010 111	57					
LD A, R	A ← R	↕	↕	X	0	X	IFF	0	•	11 101 101	ED	2	2	9		
										01 011 111	5F					
LD I, A	I ← A	•	•	X	•	X	•	•	•	11 101 101	ED	2	2	9		
										01 000 111	47					
LD R, A	R ← A	•	•	X	•	X	•	•	•	11 101 101	ED	2	2	9		
										01 001 111	4F					

Notes: r, s means any of the registers A, B, C, D, E, H, L

IFF the content of the interrupt enable flip-flop (IFF) is copied into the P/V flag

Flag Notation: • = flag not affected, 0 = flag reset, 1 = flag set, X = flag is unknown,

↕ = flag is affected according to the result of the operation.

16-BIT LOAD GROUP

Mnemonic	Symbolic Operation	Flags								Op-Code				No. of Bytes	No. of M Cycles	No. of T States	Comments	
		S	Z		H		P/V	N	C	76	543	210	Hex					
LD dd, nn	dd ← nn	•	•	X	•	X	•	•	•	00	dd0	001		3	3	10	dd	Pair
										←	n	→					00	BC
										←	n	→					01	DE
LD IX, nn	IX ← nn	•	•	X	•	X	•	•	•	11	011	101	DD	4	4	14	10	HL
										00	100	001	21				11	SP
										←	n	→						
										←	n	→						
LD IY, nn	IY ← nn	•	•	X	•	X	•	•	•	11	111	101	FD	4	4	14		
										00	100	001	21					
										←	n	→						
										←	n	→						
LD HL, (nn)	H ← (nn+1)	•	•	X	•	X	•	•	•	00	101	010	2A	3	5	16		
	L ← (nn)									←	n	→						
										←	n	→						
LD dd, (nn)	dd$_H$ ← (nn+1)	•	•	X	•	X	•	•	•	11	101	101	ED	4	6	20		
	dd$_L$ ← (nn)									01	dd1	011						
										←	n	→						
										←	n	→						
LD IX, (nn)	IX$_H$ ← (nn+1)	•	•	X	•	X	•	•	•	11	011	101	DD	4	6	20		
	IX$_L$ ← (nn)									00	101	010	2A					
										←	n	→						
										←	n	→						
LD IY, (nn)	IY$_H$ ← (nn+1)	•	•	X	•	X	•	•	•	11	111	101	FD	4	6	20		
	IY$_L$ ← (nn)									00	101	010	2A					
										←	n	→						
										←	n	→						
LD (nn), HL	(nn+1) ← H	•	•	X	•	X	•	•	•	00	100	010	22	3	5	16		
	(nn) ← L									←	n	→						
										←	n	→						
LD (nn), dd	(nn+1) ← dd$_H$	•	•	X	•	X	•	•	•	11	101	101	ED	4	6	20		
	(nn) ← dd$_L$									01	dd0	011						
										←	n	→						
										←	n	→						
LD (nn), IX	(nn+1) ← IX$_H$	•	•	X	•	X	•	•	•	11	011	101	DD	4	6	20		
	(nn) ← IX$_L$									00	100	010	22					
										←	n	→						
										←	n	→						
LD (nn), IY	(nn+1) ← IY$_H$	•	•	X	•	X	•	•	•	11	111	101	FD	4	6	20		
	(nn) ← IY$_L$									00	100	010	22					
										←	n	→						
										←	n	→						
LD SP, HL	SP ← HL	•	•	X	•	X	•	•	•	11	111	001	F9	1	1	6		
LD SP, IX	SP ← IX	•	•	X	•	X	•	•	•	11	011	101	DD	2	2	10		
										11	111	001	F9					
LD SP, IY	SP ← IY	•	•	X	•	X	•	•	•	11	111	101	FD	2	2	10		
										11	111	001	F9				qq	Pair
PUSH qq	(SP-2) ← qq$_L$	•	•	X	•	X	•	•	•	11	qq0	101		1	3	11	00	BC
	(SP-1) ← qq$_H$																01	DE
PUSH IX	(SP-2) ← IX$_L$	•	•	X	•	X	•	•	•	11	011	101	DD	2	4	15	10	HL
	(SP-1) ← IX$_H$									11	100	101	E5				11	AF
PUSH IY	(SP-2) ← IY$_L$	•	•	X	•	X	•	•	•	11	111	101	FD	2	4	15		
	(SP-1) ← IY$_H$									11	100	101	E5					
POP qq	qq$_H$ ← (SP+1)	•	•	X	•	X	•	•	•	11	qq0	001		1	3	10		
	qq$_L$ ← (SP)																	
POP IX	IX$_H$ ← (SP+1)	•	•	X	•	X	•	•	•	11	011	101	DD	2	4	14		
	IX$_L$ ← (SP)									11	100	001	E1					
POP IY	IY$_H$ ← (SP+1)	•	•	X	•	X	•	•	•	11	111	101	FD	2	4	14		
	IY$_L$ ← (SP)									11	100	001	E1					

Notes: dd is any of the register pairs BC, DE, HL, SP
        qq is any of the register pairs AF, BC, DE, HL
        (PAIR)$_H$, (PAIR)$_L$ refer to high order and low order eight bits of the register pair respectively.
        e.g. BC$_L$ = C, AF$_H$ = A

Flag Notation: • = flag not affected, 0 = flag reset, 1 = flag set, X = flag is unknown,
                ↕ flag is affected according to the result of the operation.

EXCHANGE GROUP AND BLOCK TRANSFER AND SEARCH GROUP

Mnemonic	Symbolic Operation	S	Z		H		P/V	N	C	76 543 210	Hex	No. of Bytes	No. of M Cycles	No. of T States	Comments
EX DE, HL	DE↔HL	•	•	X	•	X	•	•	•	11 101 011	EB	1	1	4	
EX AF, AF'	AF↔AF'	•	•	X	•	X	•	•	•	00 001 000	08	1	1	4	
EXX	BC↔BC' DE↔DE' HL↔HL'	•	•	X	•	X	•	•	•	11 011 001	D9	1	1	4	Register bank and auxiliary register bank exchange
EX (SP), HL	H ↔(SP+1) L ↔(SP)	•	•	X	•	X	•	•	•	11 100 011	E3	1	5	19	
EX (SP), IX	IX$_H$↔(SP+1) IX$_L$↔(SP)	•	•	X	•	X	•	•	•	11 011 101 11 100 011	DD E3	2	6	23	
EX (SP), IY	IY$_H$↔(SP+1) IY$_L$↔(SP)	•	•	X	•	X	•	•	•	11 111 101 11 100 011	FD E3	2	6	23	
LDI	(DE)←(HL) DE ← DE+1 HL ← HL+1 BC ← BC-1	•	•	X	0	X	① ↕	0	•	11 101 101 10 100 000	ED A0	2	4	16	Load (HL) into (DE), increment the pointers and decrement the byte counter (BC)
LDIR	(DE)←(HL) DE ← DE+1 HL ← HL+1 BC ← BC-1 Repeat until BC = 0	•	•	X	0	X	0	0	•	11 101 101 10 110 000	ED B0	2 2	5 4	21 16	If BC ≠ 0 If BC = 0
LDD	(DE)←(HL) DE ← DE-1 HL ← HL-1 BC ← BC-1	•	•	X	0	X	① ↕	0	•	11 101 101 10 101 000	ED A8	2	4	16	
LDDR	(DE)←(HL) DE ← DE-1 HL ← HL-1 BC ← BC-1 Repeat until BC = 0	•	•	X	0	X	0	0	•	11 101 101 10 111 000	ED B8	2 2	5 4	21 16	If BC ≠ 0 If BC = 0
CPI	A − (HL) HL ← HL+1 BC ← BC-1	↕	② ↕	X	↕	X	① ↕	1	•	11 101 101 10 100 001	ED A1	2	4	16	
CPIR	A − (HL) HL ← HL+1 BC ← BC-1 Repeat until A = (HL) or BC = 0	↕	② ↕	X	↕	X	① ↕	1	•	11 101 101 10 110 001	ED B1	2 2	5 4	21 16	If BC ≠ 0 and A ≠ (HL) If BC = 0 or A = (HL)
CPD	A − (HL) HL ← HL-1 BC ← BC-1	↕	② ↕	X	↕	X	① ↕	1	•	11 101 101 10 101 001	ED A9	2	4	16	
CPDR	A − (HL) HL ← HL-1 BC ← BC-1 Repeat until A = (HL) or BC = 0	↕	② ↕	X	↕	X	① ↕	1	•	11 101 101 10 111 001	ED B9	2 2	5 4	21 16	If BC ≠ 0 and A ≠ (HL) If BC = 0 or A = (HL)

Notes: ① P/V flag is 0 if the result of BC-1 = 0, otherwise P/V = 1
② Z flag is 1 if A = (HL), otherwise Z = 0.

Flag Notation: • = flag not affected, 0 = flag reset, 1 = flag set, X = flag is unknown,
↕ = flag is affected according to the result of the operation.

8-BIT ARITHMETIC AND LOGICAL GROUP

Mnemonic	Symbolic Operation	Flags								Op-Code			No. of Bytes	No.of M Cycles	No.of T States	Comments	
		S	Z		H		P/V	N	C	76 543 210	Hex						
ADD A, r	A ← A + r	↕	↕	X	↕	X	V	0	↕	10 [000] r			1	1	4	r	Reg.
ADD A, n	A ← A + n	↕	↕	X	↕	X	V	0	↕	11 [000] 110			2	2	7	000	B
										← n →						001	C
																010	D
ADD A, (HL)	A ← A+(HL)	↕	↕	X	↕	X	V	0	↕	10 [000] 110			1	2	7	011	E
ADD A, (IX+d)	A ← A+(IX+d)	↕	↕	X	↕	X	V	0	↕	11 011 101	DD		3	5	19	100	H
										10 [000] 110						101	L
										← d →						111	A
ADD A, (IY+d)	A ← A+(IY+d)	↕	↕	X	↕	X	V	0	↕	11 111 101	FD		3	5	19		
										10 [000] 110							
										← d →							
ADC A, s	A ← A+s+CY	↕	↕	X	↕	X	V	0	↕	[001]						s is any of r, n,	
SUB s	A ← A · s	↕	↕	X	↕	X	V	1	↕	[010]						(HL), (IX+d),	
SBC A, s	A ← A · s · CY	↕	↕	X	↕	X	V	1	↕	[011]						(IY+d) as shown for	
AND s	A ← A ∧ s	↕	↕	X	1	X	P	0	0	[100]						ADD instruction.	
OR s	A ← A ∨ s	↕	↕	X	0	X	P	0	0	[110]						The indicated bits	
XOR s	A ← A ⊕ s	↕	↕	X	0	X	P	0	0	[101]						replace the [000] in	
CP s	A · s	↕	↕	X	↕	X	V	1	↕	[111]						the ADD set above.	
INC r	r ← r + 1	↕	↕	X	↕	X	V	0	●	00 r [100]			1	1	4		
INC (HL)	(HL)←(HL)+1	↕	↕	X	↕	X	V	0	●	00 110 [100]			1	3	11		
INC (IX+d)	(IX+d) ←	↕	↕	X	↕	X	V	0	●	11 011 101	DD		3	6	23		
	(IX+d)+1									00 110 [100]							
										← d →							
INC (IY+d)	(IY+d) ←	↕	↕	X	↕	X	V	0	●	11 111 101	FD		3	6	23		
	(IY+d)+1									00 110 [100]							
										← d →							
DEC s	s ← s · 1	↕	↕	X	↕	X	V	1	●	[101]						s is any of r, (HL), (IX+d), (IY+d) as shown for INC. DEC same format and states as INC. Replace [100] with [101] in OP Code.	

Notes: The V symbol in the P/V flag column indicates that the P/V flag contains the overflow of the result of the operation. Similarly the P symbol indicates parity. V = 1 means overflow, V = 0 means not overflow, P = 1 means parity of the result is even, P = 0 means parity of the result is odd.

Flag Notation: ● = flag not affected, 0 = flag reset, 1 = flag set, X = flag is unknown.
↕ = flag is affected according to the result of the operation.

256

GENERAL PURPOSE ARITHMETIC AND CPU CONTROL GROUPS

Mnemonic	Symbolic Operation	S	Z		H		P/V	N	C	76 543 210	Hex	No. of Bytes	No. of M Cycles	No. of T States	Comments
DAA	Converts acc. content into packed BCD following add or subtract with packed BCD operands	↕	↕	X	↕	X	P	●	↕	00 100 111	27	1	1	4	Decimal adjust accumulator
CPL	A ← Ā	●	●	X	1	X	●	1	●	00 101 111	2F	1	1	4	Complement accumulator (One's complement)
NEG	A ← Ā + 1	↕	↕	X	↕	X	V	1	↕	11 101 101 01 000 100	ED 44	2	2	8	Negate acc. (two's complement)
CCF	CY ← CY̅	●	●	X	X	X	●	0	↕	00 111 111	3F	1	1	4	Complement carry flag
SCF	CY ← 1	●	●	X	0	X	●	0	1	00 110 111	37	1	1	4	Set carry flag
NOP	No operation	●	●	X	●	X	●	●	●	00 000 000	00	1	1	4	
HALT	CPU halted	●	●	X	●	X	●	●	●	01 110 110	76	1	1	4	
DI*	IFF ← 0	●	●	X	●	X	●	●	●	11 110 011	F3	1	1	4	
EI*	IFF ← 1	●	●	X	●	X	●	●	●	11 111 011	FB	1	1	4	
IM 0	Set interrupt mode 0	●	●	X	●	X	●	●	●	11 101 101 01 000 110	ED 46	2	2	8	
IM 1	Set interrupt mode 1	●	●	X	●	X	●	●	●	11 101 101 01 010 110	ED 56	2	2	8	
IM 2	Set interrupt mode 2	●	●	X	●	X	●	●	●	11 101 101 01 011 110	ED 5E	2	2	8	

Notes: IFF indicates the interrupt enable flip-flop
CY indicates the carry flip-flop.

Flag Notation: ● = flag not affected, 0 = flag reset, 1 = flag set, X = flag is unknown,
↕ = flag is affected according to the result of the operation.

*Interrupts are not sampled at the end of EI or DI

16-BIT ARITHMETIC GROUP

Mnemonic	Symbolic Operation	Flags								Op-Code			No. of Bytes	No.of M Cycles	No.of T States	Comments	
		S	Z		H		P/V	N	C	76 543 210	Hex						
ADD HL, ss	HL ← HL+ss	•	•	X	X	X	•	0	↕	00 ss1 001			1	3	11	ss	Reg.
																00	BC
ADC HL, ss	HL ← HL+ss+CY	↕	↕	X	X	X	V	0	↕	11 101 101	ED		2	4	15	01	DE
										01 ss1 010						10	HL
																11	SP
SBC HL, ss	HL ← HL-ss-CY	↕	↕	X	X	X	V	1	↕	11 101 101	ED		2	4	15		
										01 ss0 010							
ADD IX, pp	IX ← IX + pp	•	•	X	X	X	•	0	↕	11 011 101	DD		2	4	15	pp	Reg.
										00 pp1 001						00	BC
																01	DE
																10	IX
																11	SP
ADD IY, rr	IY ← IY + rr	•	•	X	X	X	•	0	↕	11 111 101	FD		2	4	15	rr	Reg.
										00 rr1 001						00	BC
																01	DE
																10	IY
																11	SP
INC ss	ss ← ss + 1	•	•	X	•	X	•	•	•	00 ss0 011			1	1	6		
INC IX	IX ← IX + 1	•	•	X	•	X	•	•	•	11 011 101	DD		2	2	10		
										00 100 011	23						
INC IY	IY ← IY + 1	•	•	X	•	X	•	•	•	11 111 101	FD		2	2	10		
										00 100 011	23						
DEC ss	ss ← ss - 1	•	•	X	•	X	•	•	•	00 ss1 011			1	1	6		
DEC IX	IX ← IX - 1	•	•	X	•	X	•	•	•	11 011 101	DD		2	2	10		
										00 101 011	2B						
DEC IY	IY ← IY - 1	•	•	X	•	X	•	•	•	11 111 101	FD		2	2	10		
										00 101 011	2B						

Notes: ss is any of the register pairs BC, DE, HL, SP
pp is any of the register pairs BC, DE, IX, SP
rr is any of the register pairs BC, DE, IY, SP.

Flag Notation: • = flag not affected, 0 = flag reset, 1 = flag set, X = flag is unknown.
↕ = flag is affected according to the result of the operation.

Z80 INSTRUCTION SET

ROTATE AND SHIFT GROUP

Mnemonic	Symbolic Operation	S	Z		H		P/V	N	C	76 543 210	Hex	No.of Bytes	No.of M Cycles	No.of T States	Comments
RLCA	[CY]←[7←0]← A	•	•		X	0	X	•	↕	00 000 111	07	1	1	4	Rotate left circular accumulator
RLA	[CY]←[7←0]← A	•	•		X	0	X	•	↕	00 010 111	17	1	1	4	Rotate left accumulator
RRCA	→[7→0]→[CY] A	•	•		X	0	X	•	↕	00 001 111	0F	1	1	4	Rotate right circular accumulator
RRA	→[7→0]→[CY] A	•	•		X	0	X	•	↕	00 011 111	1F	1	1	4	Rotate right accumulator
RLC r		↕	↕		X	0	X	P	0	↕	11 001 011 / 00 [000] r	CB	2	2	8	Rotate left circular register r
RLC (HL)		↕	↕		X	0	X	P	0	↕	11 001 011 / 00 [000] 110	CB	2	4	15	r Reg. / 000 B / 001 C
RLC (IX+d)	[CY]←[7←0]← r,(HL),(IX+d),(IY+d)	↕	↕		X	0	X	P	0	↕	11 011 101 / 11 001 011 / - d - / 00 [000] 110	DD CB	4	6	23	010 D / 011 E / 100 H
RLC (IY+d)		↕	↕		X	0	X	P	0	↕	11 111 101 / 11 001 011 / - d - / 00 [000] 110	FD CB	4	6	23	101 L / 111 A
RL s	[CY]←[7←0]← s ≡ r,(HL),(IX+d),(IY+d)	↕	↕		X	0	X	P	0	↕	[010]					Instruction format and states are as shown for RLC's. To form new Op-Code replace [000] of RLC's with shown code
RRC s	→[7→0]→[CY] s ≡ r,(HL),(IX+d),(IY+d)	↕	↕		X	0	X	P	0	↕	[001]					
RR s	→[7→0]→[CY] s ≡ r,(HL),(IX+d),(IY+d)	↕	↕		X	0	X	P	0	↕	[011]					
SLA s	[CY]←[7←0]←0 s ≡ r,(HL),(IX+d),(IY+d)	↕	↕		X	0	X	P	0	↕	[100]					
SRA s	→[7→0]→[CY] s ≡ r,(HL),(IX+d),(IY+d)	↕	↕		X	0	X	P	0	↕	[101]					
SRL s	0→[7→0]→[CY] s ≡ r,(HL),(IX+d),(IY+d)	↕	↕		X	0	X	P	0	↕	[111]					
RLD	A [7-4][3-0] [7-4][3-0](HL)	↕	↕		X	0	X	P	0	•	11 101 101 / 01 101 111	ED 6F	2	5	18	Rotate digit left and right between the accumulator and location (HL).
RRD	A [7-4][3-0] [7-4][3-0](HL)	↕	↕		X	0	X	P	0	•	11 101 101 / 01 100 111	ED 67	2	5	18	The content of the upper half of the accumulator is unaffected

Flag Notation: • = flag not affected, 0 = flag reset, 1 = flag set, X = flag is unknown,
↕ = flag is affected according to the result of the operation.

Z80 INSTRUCTION SET

BIT SET, RESET AND TEST GROUP

Mnemonic	Symbolic Operation	Flags								Op-Code			No. of Bytes	No.of M Cycles	No.of T States	Comments	
		S	Z		H		P/V	N	C	76 543 210	Hex						
BIT b, r	$Z - \overline{r}_b$	X	↕	X	1	X	X	0	•	11 001 011 01 b r	CB		2	2	8	r	Reg.
																000	B
BIT b, (HL)	$Z - \overline{(HL)}_b$	X	↕	X	1	X	X	0	•	11 001 011 01 b 110	CB		2	3	12	001	C
																010	D
BIT b, (IX+d)_b	$Z - \overline{(IX+d)}_b$	X	↕	X	1	X	X	0	•	11 011 101 11 001 011 - d - 01 b 110	DD CB		4	5	20	011 100 101 111	E H L A
																b	Bit Tested
BIT b, (IY+d)_b	$Z - \overline{(IY+d)}_b$	X	↕	X	1	X	X	0	•	11 111 101 11 001 011 - d - 01 b 110	FD CB		4	5	20	000 001 010 011 100 101 110 111	0 1 2 3 4 5 6 7
SET b, r	$r_b - 1$	•	•	X	•	X	•	•	•	11 001 011 [11] b r	CB		2	2	8		
SET b, (HL)	$(HL)_b - 1$	•	•	X	•	X	•	•	•	11 001 011 [11] b 110	CB		2	4	15		
SET b, (IX+d)	$(IX+d)_b - 1$	•	•	X	•	X	•	•	•	11 011 101 11 001 011 - d - [11] b 110	DD CB		4	6	23		
SET b, (IY+d)	$(IY+d)_b - 1$	•	•	X	•	X	•	•	•	11 111 101 11 001 011 - d - [11] b 110	FD CB		4	6	23		
RES b, s	$s_b - 0$ $s \equiv r, (HL),$ $(IX+d),$ $(IY+d)$	•	•	X	•	X	•	•	•	[10]						To form new O_p-Code replace [11] of SET b, s with [10]. Flags and time states for SET instruction	

Notes: The notation s_b indicates bit b (0 to 7) or location s.

Flag Notation: • = flag not affected, 0 = flag reset, 1 = flag set, X = flag is unknown,
↕ = flag is affected according to the result of the operation.

Z80 INSTRUCTION SET

JUMP GROUP

Mnemonic	Symbolic Operation	Flags								Op-Code			No. of Bytes	No.of M Cycles	No.of T States	Comments
		S	Z		H		P/V	N	C	76 543 210	Hex				
JP nn	PC ← nn	•	•	X	•	X	•	•	•	11 000 011 ← n → ← n →	C3	3	3	10	
JP cc, nn	If condition cc is true PC ← nn, otherwise continue	•	•	X	•	X	•	•	•	11 cc 010 ← n → ← n →		3	3	10	cc | Condition 000 | NZ non zero 001 | Z zero 010 | NC non carry 011 | C carry 100 | PO parity odd 101 | PE parity even 110 | P sign positive 111 | M sign negative
JR e	PC ← PC + e	•	•	X	•	X	•	•	•	00 011 000 ← e-2 →	18	2	3	12	
JR C, e	If C = 0, continue If C = 1, PC ← PC+e	•	•	X	•	X	•	•	•	00 111 000 ← e-2 →	38	2 2	2 3	7 12	If condition not met If condition is met
JR NC, e	If C = 1, continue If C = 0, PC ← PC+e	•	•	X	•	X	•	•	•	00 110 000 ← e-2 →	30	2 2	2 3	7 12	If condition not met If condition is met
JR Z, e	If Z = 0 continue If Z = 1, PC ← PC+e	•	•	X	•	X	•	•	•	00 101 000 ← e-2 →	28	2 2	2 3	7 12	If condition not met If condition is met
JR NZ, e	If Z = 1, continue If Z = 0, PC ← PC+e	•	•	X	•	X	•	•	•	00 100 000 ← e-2 →	20	2 2	2 3	7 12	If condition not If condition is met
JP (HL)	PC ← HL	•	•	X	•	X	•	•	•	11 101 001	E9	1	1	4	
JP (IX)	PC ← IX	•	•	X	•	X	•	•	•	11 011 101 11 101 001	DD E9	2	2	8	
JP (IY)	PC ← IY	•	•	X	•	X	•	•	•	11 111 101 11 101 001	FD E9	2	2	8	
DJNZ, e	B ← B-1 If B = 0, continue If B ≠ 0, PC ← PC+e	•	•	X	•	X	•	•	•	00 010 000 ← e-2 →	10	2 2	2 3	8 13	If B = 0 If B ≠ 0

Notes: e represents the extension in the relative addressing mode.

e is a signed two's complement number in the range <126, 129>

e-2 in the op-code provides an effective address of pc+e as PC is incremented by 2 prior to the addition of e.

Flag Notation: • = flag not affected, 0 = flag reset, 1 = flag set, X = flag is unknown,
‡ = flag is affected according to the result of the operation.

CALL AND RETURN GROUP

Mnemonic	Symbolic Operation	Flags								Op-Code		No. of Bytes	No.of M Cycles	No.of T States	Comments
		S	Z		H		P/V	N	C	76 543 210	Hex				
CALL nn	$(SP-1) \leftarrow PC_H$ $(SP-2) \leftarrow PC_L$ $PC \leftarrow nn$	•	•	X	•	X	•	•	•	11 001 101 ← n → ← n →	CD	3	5	17	
CALL cc, nn	If condition cc is false continue, otherwise same as CALL nn	•	•	X	•	X	•	•	•	11 cc 100 ← n → ← n →		3 3	3 5	10 17	If cc is false If cc is true
RET	$PC_L \leftarrow (SP)$ $PC_H \leftarrow (SP+1)$	•	•	X	•	X	•	•	•	11 001 001	C9	1	3	10	
RET cc	If condition cc is false continue, otherwise same as RET	•	•	X	•	X	•	•	•	11 cc 000		1 1	1 3	5 11	If cc is false If cc is true
RETI	Return from interrupt	•	•	X	•	X	•	•	•	11 101 101 01 001 101	ED 4D	2	4	14	
RETN[1]	Return from non maskable interrupt	•	•	X	•	X	•	•	•	11 101 101 01 000 101	ED 45	2	4	14	
RST p	$(SP-1) \leftarrow PC_H$ $(SP-2) \leftarrow PC_L$ $PC_H \leftarrow 0$ $PC_L \leftarrow p$	•	•	X	•	X	•	•	•	11 t 111		1	3	11	

cc	Condition	
000	NZ	non zero
001	Z	zero
010	NC	non carry
011	C	carry
100	PO	parity odd
101	PE	parity even
110	P	sign positive
111	M	sign negative

t	p
000	00H
001	08H
010	10H
011	18H
100	20H
101	28H
110	30H
111	38H

[1] RETN loads $IFF_2 \leftarrow IFF_1$

Flag Notation: • = flag not affected, 0 = flag reset, 1 = flag set, X = flag is unknown,
↕ = flag is affected according to the result of the operation.

INPUT AND OUTPUT GROUP

Mnemonic	Symbolic Operation	S	Z		H		P/V	N	C	76 543 210	Hex	No.of Bytes	No.of M Cycles	No.of T States	Comments
IN A, (n)	A ← (n)	•	•	X	•	X	•	•	•	11 011 011 ← n →	DB	2	3	11	n to $A_0 \sim A_7$ Acc to $A_8 \sim A_{15}$
IN r, (C)	r ← (C) if r = 110 only the flags will be affected	‡	‡	X	‡	X	P	0	•	11 101 101 01 r 000	ED	2	3	12	C to $A_0 \sim A_7$ B to $A_8 \sim A_{15}$
INI	(HL) ← (C) B ← B-1 HL ← HL+1	X	‡ ①	X	X	X	X	1	X	11 101 101 10 100 010	ED A2	2	4	16	C to $A_0 \sim A_7$ B to $A_8 \sim A_{15}$
INIR	(HL) ← (C) B ← B-1 HL ← HL+1 Repeat until B = 0	X	1	X	X	X	X	1	X	11 101 101 10 110 010	ED B2	2 2	5 (If B≠0) 4 (If B=0)	21 16	C to $A_0 \sim A_7$ B to $A_8 \sim A_{15}$
IND	(HL) ← (C) B ← B-1 HL ← HL-1	X	‡ ①	X	X	X	X	1	X	11 101 101 10 101 010	ED AA	2	4	16	C to $A_0 \sim A_7$ B to $A_8 \sim A_{15}$
INDR	(HL) ← (C) B ← B-1 HL ← HL-1 Repeat until B = 0	X	1	X	X	X	X	1	X	11 101 101 10 111 010	ED BA	2 2	5 (If B≠0) 4 (If B=0)	21 16	C to $A_0 \sim A_7$ B to $A_8 \sim A_{15}$
OUT (n), A	(n) ← A	•	•	X	•	X	•	•	•	11 010 011	D3	2	3	11	n to $A_0 \sim A_7$ Acc to $A_8 \sim A_{15}$
OUT (C), r	(C) ← r	•	•	X	•	X	•	•	•	11 101 101 01 r 001	ED	2	3	12	C to $A_0 \sim A_7$ B to $A_8 \sim A_{15}$
OUTI	B ← B-1 (C) ← (HL) HL ← HL+1	X	‡ ①	X	X	X	X	1	X	11 101 101 10 100 011	ED A3	2	4	16	C to $A_0 \sim A_7$ B to $A_8 \sim A_{15}$
OTIR	B ← B-1 (C) ← (HL) HL ← HL+1 Repeat until B = 0	X	1	X	X	X	X	1	X	11 101 101 10 110 011	ED B3	2 2	5 (If B≠0) 4 (If B=0)	21 16	C to $A_0 \sim A_7$ B to $A_8 \sim A_{15}$
OUTD	(C) ← (HL) B ← B-1 HL ← HL-1	X	‡ ①	X	X	X	X	1	X	11 101 101 10 101 011	ED AB	2	4	16	C to $A_0 \sim A_7$ B to $A_8 \sim A_{15}$
OTDR	(C) ← (HL) B ← B-1 HL ← HL-1 Repeat until B = 0	X	1	X	X	X	X	1	X	11 101 101 10 111 011	ED BB	2 2	5 (If B≠0) 4 (If B=0)	21 16	C to $A_0 \sim A_7$ B to $A_8 \sim A_{15}$

Notes: ① If the result of B - 1 is zero the Z flag is set, otherwise it is reset.

Flag Notation: • = flag not affected, 0 = flag reset, 1 = flag set, X = flag is unknown,
‡ = flag is affected according to the result of the operation.

MC 6800 INSTRUCTION SET

ACCUMULATOR AND MEMORY INSTRUCTIONS

OPERATIONS	MNEMONIC	IMMED OP	~	=	DIRECT OP	~	=	INDEX OP	~	=	EXTND OP	~	=	IMPLIED OP	~	=	BOOLEAN/ARITHMETIC OPERATION (All register labels refer to contents)	H	I	N	Z	V	C
Add	ADDA	8B	2	2	9B	3	2	AB	5	2	BB	4	3				A + M → A	↕	•	↕	↕	↕	↕
	ADDB	CB	2	2	DB	3	2	EB	5	2	FB	4	3				B + M → B	↕	•	↕	↕	↕	↕
Add Acmltrs	ABA													1B	2	1	A + B → A	↕	•	↕	↕	↕	↕
Add with Carry	ADCA	89	2	2	99	3	2	A9	5	2	B9	4	3				A + M + C → A	↕	•	↕	↕	↕	↕
	ADCB	C9	2	2	D9	3	2	E9	5	2	F9	4	3				B + M + C → B	↕	•	↕	↕	↕	↕
And	ANDA	84	2	2	94	3	2	A4	5	2	B4	4	3				A • M → A	•	•	↕	↕	R	•
	ANDB	C4	2	2	D4	3	2	E4	5	2	F4	4	3				B • M → B	•	•	↕	↕	R	•
Bit Test	BITA	85	2	2	95	3	2	A5	5	2	B5	4	3				A • M	•	•	↕	↕	R	•
	BITB	C5	2	2	D5	3	2	E5	5	2	F5	4	3				B • M	•	•	↕	↕	R	•
Clear	CLR							6F	7	2	7F	6	3				00 → M	•	•	R	S	R	R
	CLRA													4F	2	1	00 → A	•	•	R	S	R	R
	CLRB													5F	2	1	00 → B	•	•	R	S	R	R
Compare	CMPA	81	2	2	91	3	2	A1	5	2	B1	4	3				A - M	•	•	↕	↕	↕	↕
	CMPB	C1	2	2	D1	3	2	E1	5	2	F1	4	3				B - M	•	•	↕	↕	↕	↕
Compare Acmltrs	CBA													11	2	1	A - B	•	•	↕	↕	↕	↕
Complement, 1's	COM							63	7	2	73	6	3				M̄ → M	•	•	↕	↕	R	S
	COMA													43	2	1	Ā → A	•	•	↕	↕	R	S
	COMB													53	2	1	B̄ → B	•	•	↕	↕	R	S
Complement, 2's (Negate)	NEG							60	7	2	70	6	3				00 - M → M	•	•	↕	↕	①	②
	NEGA													40	2	1	00 - A → A	•	•	↕	↕	①	②
	NEGB													50	2	1	00 - B → B	•	•	↕	↕	①	②
Decimal Adjust, A	DAA													19	2	1	Converts Binary Add. of BCD Characters into BCD Format	•	•	↕	↕	:	③
Decrement	DEC							6A	7	2	7A	6	3				M - 1 → M	•	•	↕	↕	④	•
	DECA													4A	2	1	A - 1 → A	•	•	↕	↕	④	•
	DECB													5A	2	1	B - 1 → B	•	•	↕	↕	④	•
Exclusive OR	EORA	88	2	2	98	3	2	A8	5	2	B8	4	3				A ⊕ M → A	•	•	↕	↕	R	•
	EORB	C8	2	2	D8	3	2	E8	5	2	F8	4	3				B ⊕ M → B	•	•	↕	↕	R	•
Increment	INC							6C	7	2	7C	6	3				M + 1 → M	•	•	↕	↕	⑤	•
	INCA													4C	2	1	A + 1 → A	•	•	↕	↕	⑤	•
	INCB													5C	2	1	B + 1 → B	•	•	↕	↕	⑤	•
Load Acmltr	LDAA	86	2	2	96	3	2	A6	5	2	B6	4	3				M → A	•	•	↕	↕	R	•
	LDAB	C6	2	2	D6	3	2	E6	5	2	F6	4	3				M → B	•	•	↕	↕	R	•
Or, Inclusive	ORAA	8A	2	2	9A	3	2	AA	5	2	BA	4	3				A + M → A	•	•	↕	↕	R	•
	ORAB	CA	2	2	DA	3	2	EA	5	2	FA	4	3				B + M → B	•	•	↕	↕	R	•
Push Data	PSHA													36	4	1	A → MSP, SP - 1 → SP	•	•	•	•	•	•
	PSHB													37	4	1	B → MSP, SP - 1 → SP	•	•	•	•	•	•
Pull Data	PULA													32	4	1	SP + 1 → SP, MSP → A	•	•	•	•	•	•
	PULB													33	4	1	SP + 1 → SP, MSP → B	•	•	•	•	•	•
Rotate Left	ROL							69	7	2	79	6	3				M ⎫	•	•	↕	↕	⑥	↕
	ROLA													49	2	1	A ⎬ ◻─← ◻◻◻◻◻◻◻◻ ←◻ C b7 ⟵ b0	•	•	↕	↕	⑥	↕
	ROLB													59	2	1	B ⎭	•	•	↕	↕	⑥	↕
Rotate Right	ROR							66	7	2	76	6	3				M ⎫	•	•	↕	↕	⑥	↕
	RORA													46	2	1	A ⎬ ◻─→ ◻◻◻◻◻◻◻◻ →◻ C b7 → b0	•	•	↕	↕	⑥	↕
	RORB													56	2	1	B ⎭	•	•	↕	↕	⑥	↕
Shift Left, Arithmetic	ASL							68	7	2	78	6	3				M ⎫	•	•	↕	↕	⑥	↕
	ASLA													48	2	1	A ⎬ ◻ ← ◻◻◻◻◻◻◻◻ ← 0 C b7 b0	•	•	↕	↕	⑥	↕
	ASLB													58	2	1	B ⎭	•	•	↕	↕	⑥	↕
Shift Right, Arithmetic	ASR							67	7	2	77	6	3				M ⎫	•	•	↕	↕	⑥	↕
	ASRA													47	2	1	A ⎬ ◻◻◻◻◻◻◻◻ → ◻ b7 b0 C	•	•	↕	↕	⑥	↕
	ASRB													57	2	1	B ⎭	•	•	↕	↕	⑥	↕
Shift Right, Logic	LSR							64	7	2	74	6	3				M ⎫	•	•	R	↕	⑥	↕
	LSRA													44	2	1	A ⎬ 0 → ◻◻◻◻◻◻◻◻ → ◻ b7 b0 C	•	•	R	↕	⑥	↕
	LSRB													54	2	1	B ⎭	•	•	R	↕	⑥	↕
Store Acmltr	STAA				97	4	2	A7	6	2	B7	5	3				A → M	•	•	↕	↕	R	•
	STAB				D7	4	2	E7	6	2	F7	5	3				B → M	•	•	↕	↕	R	•
Subtract	SUBA	80	2	2	90	3	2	A0	5	2	B0	4	3				A - M → A	•	•	↕	↕	↕	↕
	SUBB	C0	2	2	D0	3	2	E0	5	2	F0	4	3				B - M → B	•	•	↕	↕	↕	↕
Subtract Acmltrs	SBA													10	2	1	A - B → A	•	•	↕	↕	↕	↕
Subtr with Carry	SBCA	82	2	2	92	3	2	A2	5	2	B2	4	3				A - M - C → A	•	•	↕	↕	↕	↕
	SBCB	C2	2	2	D2	3	2	E2	5	2	F2	4	3				B - M - C → B	•	•	↕	↕	↕	↕
Transfer Acmltrs	TAB													16	2	1	A → B	•	•	↕	↕	R	•
	TBA													17	2	1	B → A	•	•	↕	↕	R	•
Test, Zero or Minus	TST							6D	7	2	7D	6	3				M - 00	•	•	↕	↕	R	R
	TSTA													4D	2	1	A - 00	•	•	↕	↕	R	R
	TSTB													5D	2	1	B - 00	•	•	↕	↕	R	R
																		H	I	N	Z	V	C

INDEX REGISTER AND STACK MANIPULATION INSTRUCTIONS

POINTER OPERATIONS	MNEMONIC	IMMED OP	~	=	DIRECT OP	~	=	INDEX OP	~	#	EXTND OP	~	#	IMPLIED OP	~	=	BOOLEAN/ARITHMETIC OPERATION	5 H	4 I	3 N	2 Z	1 V	0 C
Compare Index Reg	CPX	8C	3	3	9C	4	2	AC	6	2	BC	5	3				$X_H - M$, $X_L - (M+1)$	•	•	⑦	‡	⑧	•
Decrement Index Reg	DEX													09	4	1	$X - 1 \rightarrow X$	•	•	•	‡	•	•
Decrement Stack Pntr	DES													34	4	1	$SP - 1 \rightarrow SP$	•	•	•	•	•	•
Increment Index Reg	INX													08	4	1	$X + 1 \rightarrow X$	•	•	•	‡	•	•
Increment Stack Pntr	INS													31	4	1	$SP + 1 \rightarrow SP$	•	•	•	•	•	•
Load Index Reg	LDX	CE	3	3	DE	4	2	EE	6	2	FE	5	3				$M \rightarrow X_H$, $(M+1) \rightarrow X_L$	•	•	⑨	‡	R	•
Load Stack Pntr	LDS	8E	3	3	9E	4	2	AE	6	2	BE	5	3				$M \rightarrow SP_H$, $(M+1) \rightarrow SP_L$	•	•	⑨	‡	R	•
Store Index Reg	STX				DF	5	2	EF	7	2	FF	6	3				$X_H \rightarrow M$, $X_L \rightarrow (M+1)$	•	•	⑨	‡	R	•
Store Stack Pntr	STS				9F	5	2	AF	7	2	BF	6	3				$SP_H \rightarrow M$, $SP_L \rightarrow (M+1)$	•	•	⑨	‡	R	•
Indx Reg → Stack Pntr	TXS													35	4	1	$X - 1 \rightarrow SP$	•	•	•	•	•	•
Stack Pntr → Indx Reg	TSX													30	4	1	$SP + 1 \rightarrow X$	•	•	•	•	•	•

JUMP AND BRANCH INSTRUCTIONS

OPERATIONS	MNEMONIC	RELATIVE OP	~	=	INDEX OP	~	#	EXTND OP	~	#	IMPLIED OP	~	#	BRANCH TEST	5 H	4 I	3 N	2 Z	1 V	0 C
Branch Always	BRA	20	4	2										None	•	•	•	•	•	•
Branch If Carry Clear	BCC	24	4	2										$C = 0$	•	•	•	•	•	•
Branch If Carry Set	BCS	25	4	2										$C = 1$	•	•	•	•	•	•
Branch If = Zero	BEQ	27	4	2										$Z = 1$	•	•	•	•	•	•
Branch If ≥ Zero	BGE	2C	4	2										$N \oplus V = 0$	•	•	•	•	•	•
Branch If > Zero	BGT	2E	4	2										$Z + (N \oplus V) = 0$	•	•	•	•	•	•
Branch If Higher	BHI	22	4	2										$C + Z = 0$	•	•	•	•	•	•
Branch If ≤ Zero	BLE	2F	4	2										$Z + (N \oplus V) = 1$	•	•	•	•	•	•
Branch If Lower Or Same	BLS	23	4	2										$C + Z = 1$	•	•	•	•	•	•
Branch If < Zero	BLT	2D	4	2										$N \oplus V = 1$	•	•	•	•	•	•
Branch If Minus	BMI	2B	4	2										$N = 1$	•	•	•	•	•	•
Branch If Not Equal Zero	BNE	26	4	2										$Z = 0$	•	•	•	•	•	•
Branch If Overflow Clear	BVC	28	4	2										$V = 0$	•	•	•	•	•	•
Branch If Overflow Set	BVS	29	4	2										$V = 1$	•	•	•	•	•	•
Branch If Plus	BPL	2A	4	2										$N = 0$	•	•	•	•	•	•
Branch To Subroutine	BSR	8D	8	2										} See Special Operations	•	•	•	•	•	•
Jump	JMP				6E	4	2	7E	3	3				} See Special Operations	•	•	•	•	•	•
Jump To Subroutine	JSR				AD	8	2	BD	9	3				}	•	•	•	•	•	•
No Operation	NOP										01	2	1	Advances Prog. Cntr. Only	•	•	•	•	•	•
Return From Interrupt	RTI										3B	10	1		—	—	(⑩)	—	—	—
Return From Subroutine	RTS										39	5	1	}	•	•	•	•	•	•
Software Interrupt	SWI										3F	12	1	} See Special Operations	•	•	•	•	•	•
Wait for Interrupt *	WAI										3E	9	1	}	•	⑪	•	•	•	•

*WAI puts Address Bus, R/W, and Data Bus in the three state mode while VMA is held low.

CONDITION CODE REGISTER MANIPULATION INSTRUCTIONS

OPERATIONS	MNEMONIC	IMPLIED OP	~	=	BOOLEAN OPERATION	5 H	4 I	3 N	2 Z	1 V	0 C
Clear Carry	CLC	0C	2	1	$0 \rightarrow C$	•	•	•	•	•	R
Clear Interrupt Mask	CLI	0E	2	1	$0 \rightarrow I$	•	R	•	•	•	•
Clear Overflow	CLV	0A	2	1	$0 \rightarrow V$	•	•	•	•	R	•
Set Carry	SEC	0D	2	1	$1 \rightarrow C$	•	•	•	•	•	S
Set Interrupt Mask	SEI	0F	2	1	$1 \rightarrow I$	•	S	•	•	•	•
Set Overflow	SEV	0B	2	1	$1 \rightarrow V$	•	•	•	•	S	•
Acmltr A → CCR	TAP	06	2	1	$A \rightarrow CCR$	—	—	(⑫)	—	—	—
CCR → Acmltr A	TPA	07	2	1	$CCR \rightarrow A$	•	•	•	•	•	•

CONDITION CODE REGISTER NOTES: (Bit set if test is true and cleared otherwise)

1 (Bit V) Test: Result = 10000000?
2 (Bit C) Test: Result = 000000000?
3 (Bit C) Test: Decimal value of most significant BCD Character greater than nine? (Not cleared if previously set.)
4 (Bit V) Test: Operand = 10000000 prior to execution?
5 (Bit V) Test: Operand = 01111111 prior to execution?
6 (Bit V) Test: Set equal to result of N⊕C after shift has occurred

7 (Bit N) Test: Sign bit of most significant (MS) byte = 1?
8 (Bit V) Test: 2's complement overflow from subtraction of MS bytes?
9 (Bit N) Test: Result less than zero? (Bit 15 = 1)
10 (All) Load Condition Code Register from Stack. (See Special Operations)
11 (Bit I) Set when interrupt occurs. If previously set, a Non Maskable Interrupt is required to exit the wait state.
12 (All) Set according to the contents of Accumulator A.

Answers to multi-choice problems

Chapter 2 (page 41)
1 (b); 2 (c); 3 (d); 4 (a); 5 (d); 6 (c); 7 (d); 8 (a); 9 (a); 10 (b).

Chapter 3 (page 91)
1 (a); 2 (c); 3 (a); 4 (b); 5 (c); 6 (d); 7 (a); 8 (c); 9 (d); 10 (a).

Index